10 Indiana ILEARN Grade 7 Math Practice Tests

The Ultimate Test Prep Collection with Answer Explanations

Dr. A. Nazari

10 Practice Tests

Grade 7 Math — Engineered for Mastery

Welcome to **The Architect's Workshop.**

Ten practice tests. Ten floors of a tower you're about to build. Each test lays another level of understanding — from foundation to capstone.

- **Foundation (Tests 1–3):** Learn the blueprints
- **Framework (Tests 4–6):** Build structural strength
- **Finishing (Tests 7–9):** Refine under pressure
- **Capstone (Test 10):** The final inspection

Precision. Practice. Perfection.

> **"** Every great structure
> is built one floor at a time.
> Ten levels of practice makes
> your math rock-solid. **"**

✎ The Architectural Plan ✎

A 4-phase construction plan for complete mastery

Phase I: Foundation (Tests 1–3)

Untimed. Explore the test format and question types. Read every answer explanation after each test. Goal: understand the blueprint before you build.

Phase II: Framework (Tests 4–6)

Add a timer (70 minutes). Focus on the topics that gave you trouble in Phase I. Practice showing complete work and labeling units. Goal: build structural strength.

Phase III: Finishing (Tests 7–9)

Full timed conditions (60 minutes). Simulate the real exam. Review only the questions you missed — don't re-study what you already know. Goal: refine under real pressure.

Phase IV: Capstone (Test 10)

Your final inspection. Full exam conditions. This is the capstone — compare with Test 1 and measure your entire growth arc. Goal: prove your mastery.

📋 Blueprint Specifications

✎ **Multiple Choice** — select the single best answer

✎ **Multi-Select** — choose ALL that apply

✎ **Short Answer** — show your process

✎ **Open Response** — explain and justify

Engineering Principles

Precision techniques for building every solution

The D.R.A.F.T. Method

D — **Define** — What exactly is the question asking? Write it in your own words.

R — **Retrieve** — Pull out given information. List numbers, units, and relationships.

A — **Assemble** — Choose the right formula or strategy. Set up the equation or proportion.

F — **Figure** — Compute step by step. Show every operation. Label units.

T — **Test** — Does the answer make sense? Re-read the question. Verify the units.

Seven Precision Rules

1. **Read twice, solve once.** The first read gives context; the second reveals what to compute.

3. **Estimate before calculating.** A quick mental approximation catches major errors.

5. **Track your signs.** Rational number operations are the #1 error source in Grade 7.

7. **Never submit blanks.** On open response, even a partial setup can earn credit.

Timing Blueprint

Tests 1–3: **Untimed** (learn the blueprints) > Tests 4–6: **70 min** (build speed) > Tests 7–10: **60 min** (exam conditions)

 *Every great structure starts with a solid plan. Study the blueprints, learn from each test, and build something extraordinary.*

Find more at
ViewMath.com/IN-Grade7

🧰 The Drafting Toolkit 🧰

Prepare your station before every construction session

⚖ Required Equipment

Precision Pencils — *Two or more #2 pencils, kept sharp. A dull tool produces dull work.*

Quality Eraser — *Clean, smudge-free corrections. Architects revise — so will you.*

Scratch Paper — *Your drafting workspace. All calculations happen here first.*

Ruler / Protractor — *Precision tools for scale drawings, angle measures, and geometry.*

Timer — *Begin using from Phase II (Test 4) onward.*

Quiet Workspace — *A clean, well-lit station free from distractions.*

✓ Approved Materials

- ✔ Pencil and eraser
- ✔ Scratch paper (provided on test day)
- ✔ Ruler or protractor (if specified)
- ✔ Reference sheet (if provided)

🚫 Restricted Materials

- ✘ Calculator (unless specified)
- ✘ Electronic devices
- ✘ Notes or reference materials
- ✘ Communication with others

♥ Project Manager's Guide (for Parents & Guardians)

Ten practice tests provide the most comprehensive preparation available. The 4-phase structure (Foundation → Framework → Finishing → Capstone) ensures skills build progressively and durably.

Keys to a successful build:

- Space tests 2–3 days apart — never more than one per day
- Phases I–II: review answers together and discuss strategies
- Phases III–IV: let them work independently, then debrief results
- Compare Test 1 with Test 10 to celebrate the full construction arc

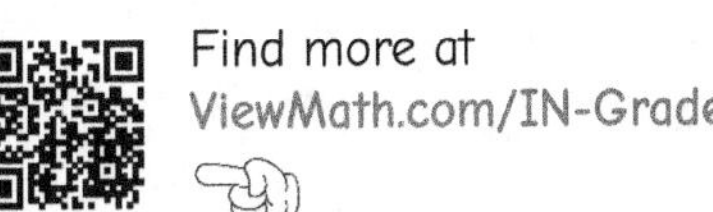

Find more at
ViewMath.com/IN-Grade7

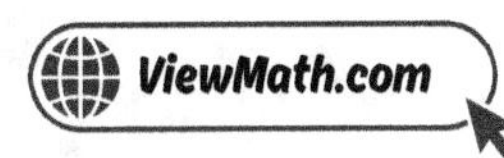

Math Reference Sheet

You may use this page during your practice tests!

Symbol	Name	What It Means	
$(\ )$	Parentheses	Do this part first.	$(3+4) \times 2 = 14$
10^3	Exponent	Multiply the base by itself that many times. $10^3 = 1,000$	
$\frac{a}{b}$	Fraction	a parts out of b equal parts; also means $a \div b$.	
$\frac{7}{3}$	Improper Fraction	Numerator $\geq$ denominator.	$\frac{7}{3} = 2\frac{1}{3}$
0.45	Decimal	A number with a decimal point.	$0.45 = \frac{45}{100}$
$> < =$	Comparison	Greater than, less than, equal to.	$0.5 > 0.35$
$(3,5)$	Ordered Pair	A point on the coordinate plane: (x, y).	

Key Formulas

- **Volume of a rectangular prism:**

  $V = l \times w \times h$

- **Order of operations:**

  Parentheses $\rightarrow$ Exponents $\rightarrow$ Multiply/Divide $\rightarrow$ Add/Subtract

- **Powers of 10:**

  $10^1 = 10 \quad 10^2 = 100$

  $10^3 = 1,000 \quad 10^4 = 10,000$

- **Fraction as division:**

  $\frac{a}{b} = a \div b$

Place Value Chart

Millions	1,000,000
Hundred-Thousands	100,000
Ten-Thousands	10,000
Thousands	1,000
Hundreds	100
Tens	10
Ones	1

Decimals

Tenths	0.1
Hundredths	0.01
Thousandths	0.001

Each place is $10\times$ the place to its right, and $\frac{1}{10}$ of the place to its left.

Key Math Vocabulary

- **Sum** — the result of addition
- **Difference** — the result of subtraction
- **Product** — the result of multiplication
- **Quotient** — the result of division
- **Remainder** — what's left over after dividing
- **Factor** — a number you multiply
- **Expression** — numbers and operations without $=$
- **Equation** — a math sentence with $=$
- **Numerator** — the top number of a fraction
- **Denominator** — the bottom number of a fraction
- **Mixed number** — a whole number + a fraction
- **Equivalent fractions** — fractions with equal value
- **Decimal** — a number written with a decimal point
- **Volume** — the space inside a 3D shape
- **Coordinate plane** — a grid with x and y axes
- **Ordered pair** — (x, y) location on the plane

Word Problem Clue Words

- **Add** $(+)$: total, altogether, combined, sum, increase, more than
- **Subtract** $(-)$: difference, how many more, fewer, remain, decrease, left
- **Multiply** $(\times)$: each, every, groups of, times, product, per, of (with fractions)
- **Divide** $(\div)$: share equally, split, each group, how many groups, quotient, per

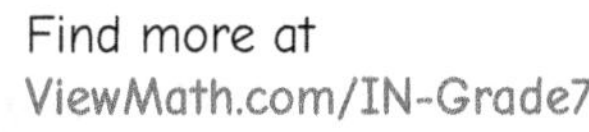

Find more at
ViewMath.com/IN-Grade7

⬛ Multiplication Table ⬛

You may use this table during your practice tests!

×	1	2	3	4	5	6	7	8	9	10	11
1	1	2	3	4	5	6	7	8	9	10	11
2	2	4	6	8	10	12	14	16	18	20	22
3	3	6	9	12	15	18	21	24	27	30	33
4	4	8	12	16	20	24	28	32	36	40	44
5	5	10	15	20	25	30	35	40	45	50	55
6	6	12	18	24	30	36	42	48	54	60	66
7	7	14	21	28	35	42	49	56	63	70	77
8	8	16	24	32	40	48	56	64	72	80	88
9	9	18	27	36	45	54	63	72	81	90	99
10	10	20	30	40	50	60	70	80	90	100	110
11	11	22	33	44	55	66	77	88	99	110	121

💡 How to Use This Table

To find **4 × 7**:

1. Find **4** in the left column (blue).
2. Find **7** in the top row (blue).
3. Follow the row and column until they meet: the answer is **28**!

ℹ Tip: You can also use this table for division! If you know 28 ÷ 4 = ?, find 28 in the 4's row. The column header gives you the answer: **7**!

🏢 Construction Log 🏢

Record each floor and watch your tower rise

Architect: ________________________________   Project Start: ________________

Phase I: Foundation — Floors 1-3

Floor 1   Date: __________   Score: _____ / _____   %: _____   ★★★★★
Floor 2   Date: __________   Score: _____ / _____   %: _____   ★★★★★
Floor 3   Date: __________   Score: _____ / _____   %: _____   ★★★★★

Phase II: Framework — Floors 4-6

Floor 4   Date: __________   Score: _____ / _____   %: _____   ★★★★★
Floor 5   Date: __________   Score: _____ / _____   %: _____   ★★★★★
Floor 6   Date: __________   Score: _____ / _____   %: _____   ★★★★★

Phase III: Finishing — Floors 7-9

Floor 7   Date: __________   Score: _____ / _____   %: _____   ★★★★★
Floor 8   Date: __________   Score: _____ / _____   %: _____   ★★★★★
Floor 9   Date: __________   Score: _____ / _____   %: _____   ★★★★★

Phase IV: Capstone — Floor 10

Floor 10   Date: __________   Score: _____ / _____   %: _____   ★★★★★

Certification Level: Apprentice (0–39%)   Technician (40–59%)   Engineer (60–79%)   **Master Architect** (80–100%)

Record your certification level after each test!

🔧 Under Construction

- ____
- ____
- ____
- ____

✅ Construction Complete

- ⭐ ____
- ⭐ ____
- ⭐ ____
- ⭐ ____

Find more at
ViewMath.com/IN-Grade7

 Table of Contents

Here's what we'll explore together!

 Let's learn and have fun!

Practice Test 1

 30 Questions

✏️ Before You Start ✏️

- ✔ **Read each question carefully** before choosing your answer.
- ✔ **Show your work** on scratch paper when you need to.
- ✔ **Skip hard questions** and come back to them later.
- ✔ **Check your answers** when you're done.
- ✔ **Take your time** — there's no rush!

 You've Got This! 

Do your best and show what you know!

1. The table shows a proportional relationship. Find k and the missing value.

x	y
3	7.5
?	15
10	25

Your Answer:

2. Two proportional lines are graphed below. Which statement is correct?

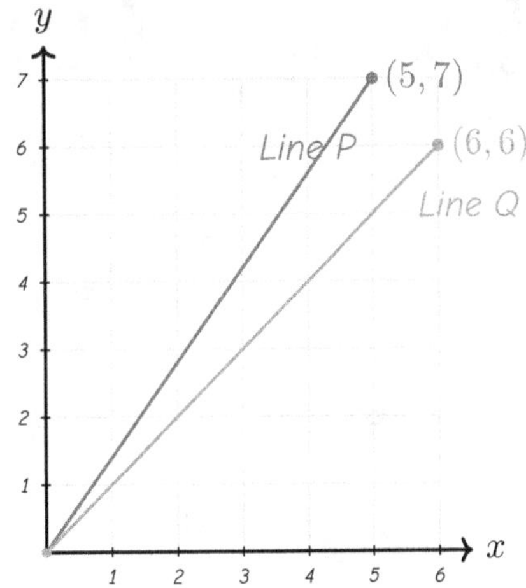

(A) Line Q has the greater unit rate because it is longer

(B) Line P has the greater unit rate because it is steeper

(C) Both lines have the same unit rate

(D) Line Q has the greater unit rate because $6 > 5$

3. The graph below compares the rates of two workers packaging boxes. Which worker is faster, and by how many boxes per hour?

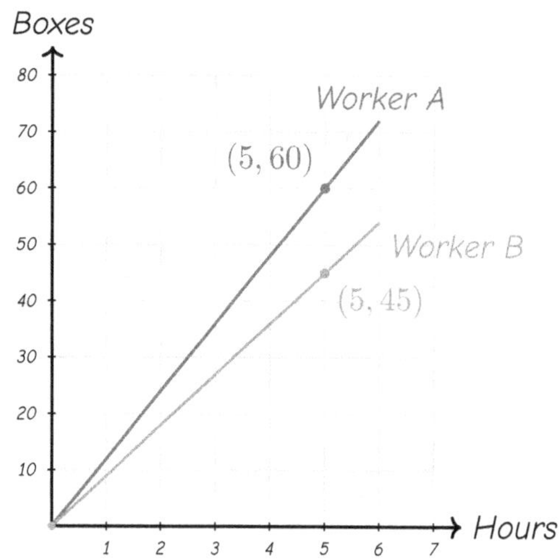

(A) Worker A, by 3 boxes/hr

(B) Worker B, by 3 boxes/hr

(C) Worker A, by 15 boxes/hr

(D) They work at the same rate

4. A school has 500 students. If 68% eat in the cafeteria, how many eat there? Use a proportion.

(A) 320

(B) 340

(C) 350

(D) 368

5. The diagram shows how a salesperson's earnings are calculated. What are the total earnings?

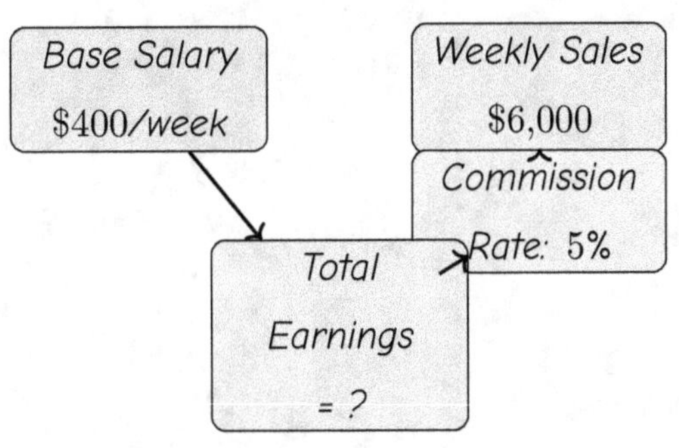

(A) $430

(B) $700

(C) $6,400

(D) $6,700

6. A jogger estimated running 5 miles but actually ran 4.8 miles. What is the percent error (rounded to the nearest tenth)?

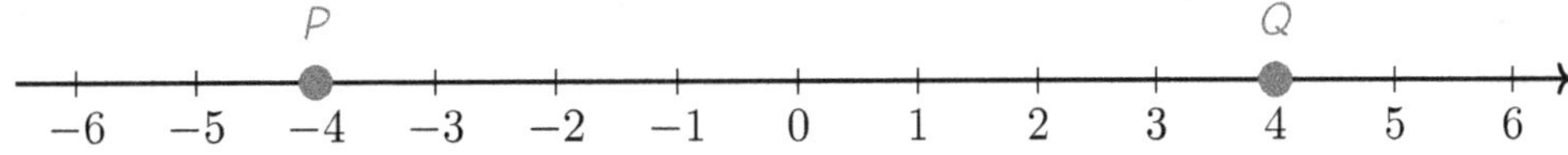

Your Answer:

7. The number line below shows two points. Which statement is correct?

(A) P and Q are the same number

(B) P and Q are opposites

(C) $|P| > |Q|$

(D) $P > Q$

8. What is $0 + (-7)$?

(A) 7

(B) 0

(C) −7

(D) −14

Find more at
ViewMath.com/IN-Grade7

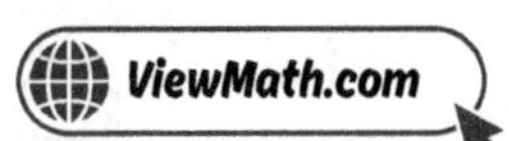

9. *Simplify* $12 - 5w + 3w - 8 + w.$

Your Answer

10. *Expand and simplify* $3(2x + 4) - 5x.$

Your Answer

11. *The thermometer shows the current temperature. The temperature was* $-3.5°F$ *this morning and rose t degrees each hour for 4 hours to reach the current reading. What is t?*

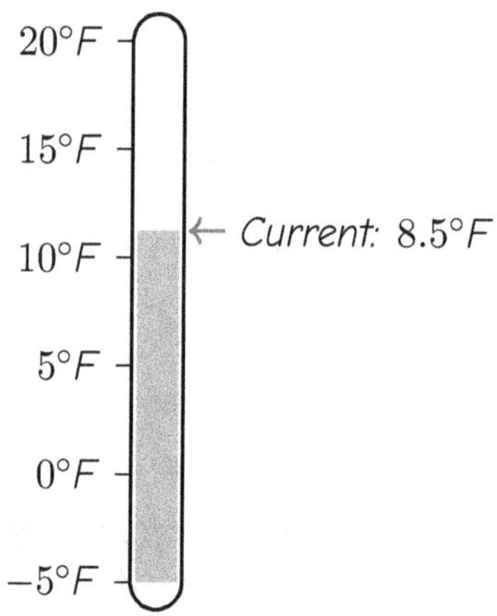

(A) $t = 3$

(B) $t = 5$

(C) $t = 1.25$

(D) $t = 2$

12. The table shows ticket prices for groups. A school has $200 to spend and must also pay a $35 bus fee. Write and solve an inequality to find the maximum number of students s who can go.

Group Size	Price per Student
1–10	$12
11–25	$9
26+	$7

Assume the group qualifies for the $9 rate.

Your Answer:

13. A phone battery is at 85% and loses 7% per hour. Write an inequality for when the battery is below 20%, and solve.

Your Answer

14. A map uses the scale 1 cm = 25 km. Two towns are 6 cm apart on the map. What is the actual distance?

(A) 31 km

(B) 125 km

(C) 150 km

(D) 175 km

15. You are given angles 45° and 45° and side 6 cm between them. How many triangles can you draw?

Your Answer

16. Two sides of a triangle are 9 cm and 12 cm with an included angle of 60°. What type of condition is this?

(A) SSS

(B) SAS

(C) ASA

(D) AAA

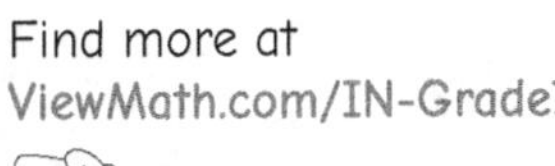
Find more at
ViewMath.com/IN-Grade7

17. Name two different cross-section shapes you can get from slicing a cube.

Your Answer:

18. Two vertical angles are $(7x + 1)°$ and $(5x + 13)°$. Find x and the angle measure.

Your Answer:

19. A figure is translated so that point $(2, 7)$ moves to $(2, 3)$. What translation was applied?

(A) Up 4

(B) Down 4

(C) Left 4

(D) Right 4

20. A circle has a diameter of 14 cm. What is its area? Use $\pi \approx \frac{22}{7}$.

(A) 44 cm^2

(B) 154 cm^2

(C) 308 cm^2

(D) 616 cm^2

21. A figure is made of a rectangle 12 cm by 5 cm with a right triangle removed. The triangle has legs 4 cm and 5 cm. What is the area of the remaining figure?

(A) 40 cm^2

(B) 50 cm^2

(C) 60 cm^2

(D) 70 cm^2

22. A triangular prism has a triangular base with base 6 cm and height 4 cm, and the prism is 10 cm long. The three rectangular faces have widths 6 cm, 5 cm, and 5 cm. What is the total surface area?

(A) 184 cm^2

(B) 160 cm^2

(C) 124 cm^2

(D) 120 cm^2

Find more at
ViewMath.com/IN-Grade7

23. A rectangular fish tank is 50 cm long, 30 cm wide, and 40 cm tall. What is the volume?

(A) $120\ cm^3$

(B) $6{,}000\ cm^3$

(C) $60{,}000\ cm^3$

(D) $600\ cm^3$

24. In a school of 600 students, a random sample of 60 found that 15 students walk to school. Predict how many students in the entire school walk to school.

(A) 100

(B) 150

(C) 200

(D) 250

25. Which measure of variability is shown directly on a box plot?

(A) Mean absolute deviation (MAD)

(B) Standard deviation

(C) Interquartile range (IQR)

(D) Mean

26. The table compares two delivery companies' shipping times (in days).

	Company X	Company Y
Mean	3.5 days	3.2 days
Median	3 days	3 days
Range	4 days	8 days
IQR	1.5 days	4 days

A customer wants the most reliable delivery time. Which company should they choose?

(A) Company Y, because its mean is lower

(B) Company X, because its IQR and range are both smaller

(C) Company Y, because its range includes more possibilities

(D) They are equally reliable

27. A double bar graph compares boys and girls in a math competition. Boys scored $80, 85, 90$ and girls scored $75, 90, 95$ over three rounds. Who had the higher average?

(A) Boys (85)

(B) Girls (86.7)

(C) They tied

(D) Cannot be determined

28. A probability model has outcomes X, Y, and Z. $P(X) = 0.15$ and $P(Y) = 0.45$. What is $P(Z)$?

Your Answer

29. A spinner has 4 equal sections $(1, 2, 3, 4)$ and a die is rolled. What is the probability that both show a 3?

(A) $\frac{1}{10}$

(B) $\frac{1}{24}$

(C) $\frac{2}{10}$

(D) $\frac{1}{4}$

30. A spinner has a 60% chance of landing on blue. In a simulation of 80 spins, how many times would you expect blue to appear?

Your Answer

 End of Practice Test 1 

Great job finishing the test!

✅ My Score

I got _____________ out of 30 questions right.

Check your answers in the **Answer Key** at the back of the book.

💡 Review any questions you missed. That's how we learn!

📊 Check Your Score Online!

Visit **ViewMath Academy** to enter your answers and see which topics you need to review. You can also explore lessons, take quizzes, track your scores, and save your progress!

viewmath.com/score/7.1.IN.16

Or go to viewmath.com/score and enter code: 7.1.IN.16

Practice Test 2

30 Questions

✏ Before You Start ✏

- ✔ **Read each question carefully** before choosing your answer.
- ✔ **Show your work** on scratch paper when you need to.
- ✔ **Skip hard questions** and come back to them later.
- ✔ **Check your answers** when you're done.
- ✔ **Take your time** — there's no rush!

★ You've Got This! ★

Do your best and show what you know!

1. The table below shows a proportional relationship between gallons of gas and miles driven. What is k and what does it mean?

Gallons (x)	Miles (y)
2	54
5	135
8	216

(A) $k = 2$; uses 2 gallons per trip

(B) $k = 27$; drives 27 miles per gallon

(C) $k = 54$; drives 54 miles on 2 gallons

(D) $k = 135$; drives 135 miles on 5 gallons

2. The graph below shows the number of birdhouses a carpenter can build over time. What is the unit rate, and how many birdhouses can be built in 10 hours?

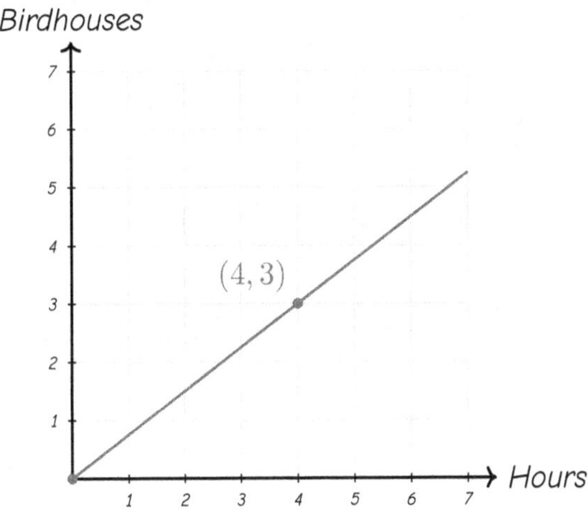

Your Answer

Find more at
ViewMath.com/IN-Grade7

3. The table shows how many batches of muffins a bakery can make and the total flour used. How many cups of flour are needed for 15 batches?

Batches	Flour (cups)
3	$7\frac{1}{2}$
6	15
9	$22\frac{1}{2}$
15	?

Your Answer:

4. Which proportion shows that 42 is 60% of some number w?

(A) $\dfrac{42}{w} = \dfrac{60}{100}$

(B) $\dfrac{w}{42} = \dfrac{60}{100}$

(C) $\dfrac{60}{42} = \dfrac{w}{100}$

(D) $\dfrac{42}{60} = \dfrac{w}{100}$

5. A bill is \$86 and you leave a \$17.20 tip. What percent tip is that?

(A) 15%

(B) 17%

(C) 18%

(D) 20%

6. Two students estimated the number of beans in a jar. The actual count was 200. Student A guessed 180; Student B guessed 220. Who had the greater percent error?

(A) Student A

(B) Student B

(C) Same percent error

(D) Cannot be determined

7. What is $|-9|$?

(A) -9

(B) 0

(C) $\frac{1}{9}$

(D) 9

8. Find the sum: $5 + (-11) + 3 + (-2)$.

Your Answer:

9. Simplify $-4x + 9 + 7x - 3$.

(A) $3x + 6$

(B) $-11x + 6$

(C) $3x - 6$

(D) $11x + 12$

10. Expand $\frac{3}{4}(8a - 12)$.

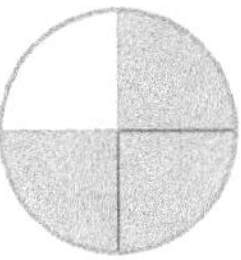

Your Answer:

11. Solve $\dfrac{5x}{2} + 3 = 18$.

(A) $x = 3$

(B) $x = 6$

(C) $x = 42$

(D) $x = 15$

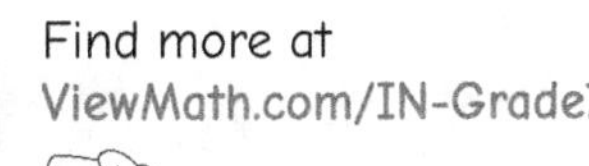

Find more at
ViewMath.com/IN-Grade7

ViewMath.com

12. Solve $8 - 3x \geq 23$.

Your Answer:

13. A student has test scores of 78, 85, and 90. She needs an average of at least 85 on four tests to earn an A. Solve the inequality to find the minimum score s on the fourth test, and describe the graph.

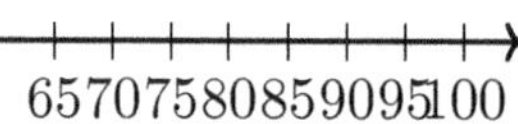

Your Answer:

14. A drawing of a room uses scale 1 cm = 3 m. The drawing shows a square room with side 4 cm. If the scale factor doubles all lengths, by what factor does the actual area change?

Your Answer:

15. Can a triangle have sides 6, 8, and 14?

(A) Yes, it is a right triangle

(B) Yes, it is an obtuse triangle

(C) No, the triangle inequality is not satisfied

(D) No, because the sides are all even numbers

16. Two angles of a triangle are 35° and 75°. What is the third angle?

(A) 60°

(B) 70°

(C) 80°

(D) 110°

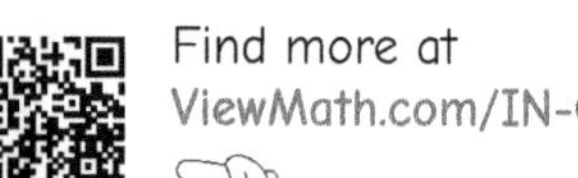
Find more at
ViewMath.com/IN-Grade7

17. You slice a cube with a vertical cut from one face to the opposite face, perpendicular to the base. What is the most likely cross-section shape?

(A) Square or rectangle

(B) Circle

(C) Triangle

(D) Hexagon

18. Two angles are supplementary. One is 25° more than the other. What are the two angles?

(A) 65° and 115°

(B) 77.5° and 102.5°

(C) 80° and 100°

(D) 75° and 105°

19. Reflect point $P(-6, 2)$ over the x-axis. What are the coordinates of P'?

Your Answer

20. A circle has an area of 78.5 cm^2. Using $\pi \approx 3.14$, what is the radius?

(A) 4 cm

(B) 5 cm

(C) 10 cm

(D) 25 cm

21. A swimming pool is shaped like a rectangle 25 m by 10 m with a semicircle of diameter 10 m at one end. What is the total area? Use $\pi \approx 3.14$

(A) 250 m^2

(B) 289.25 m^2

(C) 328.5 m^2

(D) 367.75 m^2

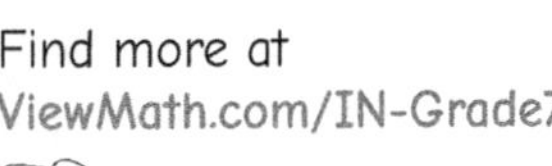
Find more at
ViewMath.com/IN-Grade7

22. A triangular prism has a right-triangle base with legs 3 cm and 4 cm. The prism is 12 cm long. The hypotenuse of the triangle is 5 cm. What is the total surface area?

Your Answer:

23. A triangular prism has a base that is a triangle with base 8 cm and height 3 cm. The prism is 10 cm long. What is the volume?

(A) 24 cm^3

(B) 80 cm^3

(C) 120 cm^3

(D) 240 cm^3

24. A factory produces 10,000 batteries per day. A sample of 200 batteries is tested, and 8 are defective. What percent of the sample is defective?

Your Answer:

25. The dot plots below show the number of push-ups completed by students in two gym classes.

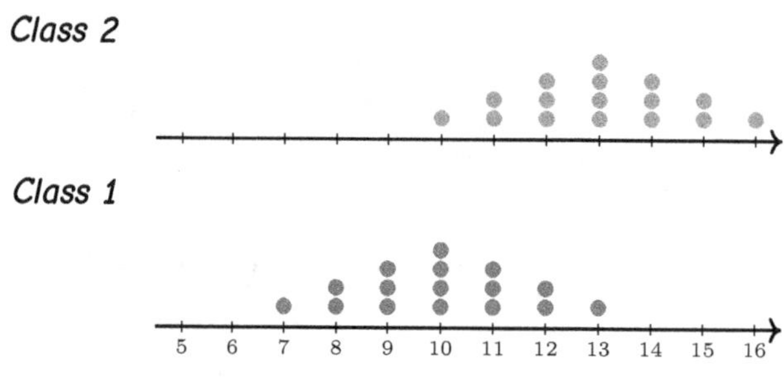

Which class has a higher center, and by approximately how many push-ups?

(A) Class 1, by about 3

(B) Class 2, by about 3

(C) Class 2, by about 6

(D) They have the same center

26. The dot plots show daily high temperatures (°F) for two cities over 10 days.

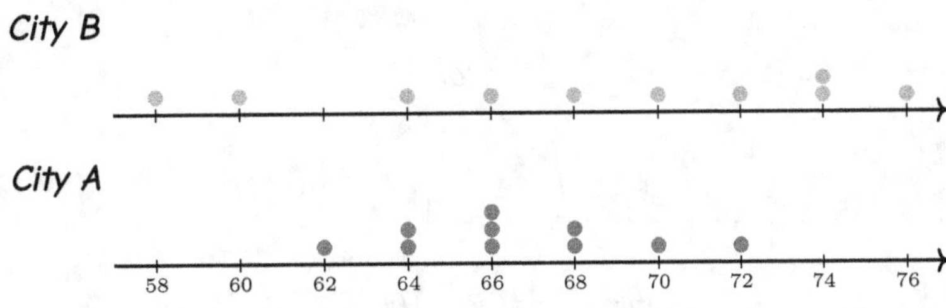

Find the mean and range for each city. Which city has more consistent temperatures?

Your Answer:

27. A frequency table lists:

(A) Only the largest and smallest values

(B) Categories and how often each occurs

(C) The mean, median, and mode

(D) Data arranged in a circle

28. A student creates a probability model for a spinner. $P(red) = 0.25$, $P(blue) = 0.25$, $P(green) = 0.25$, $P(yellow) = 0.25$. After 200 spins, the results are: red = 55, blue = 48, green = 52, yellow = 45. Is the model a good fit?

(A) No, the model is completely wrong

(B) Yes, the observed frequencies are reasonably close to 50 each

(C) No, because the results are not exactly 50 each

(D) Yes, because 200 is divisible by 4

29. The tree diagram shows the outcomes for flipping two coins. What is the probability of getting at least one tail?

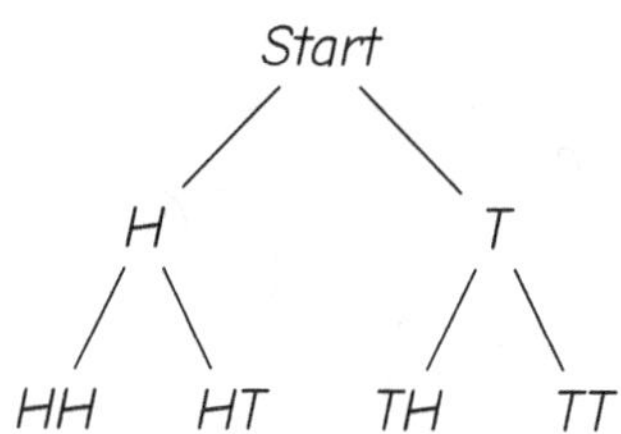

(A) $\frac{1}{4}$

(B) $\frac{1}{2}$

(C) $\frac{3}{4}$

(D) $\frac{2}{4}$

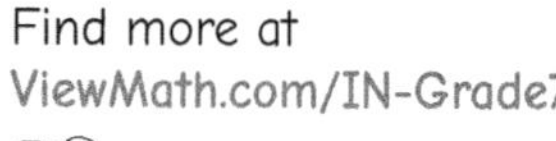

Find more at
ViewMath.com/IN-Grade7

ViewMath.com

30. The bar graph shows the results of simulating a spinner 100 times. The spinner is supposed to land on each color with equal probability. Based on the simulation, does the spinner appear fair?

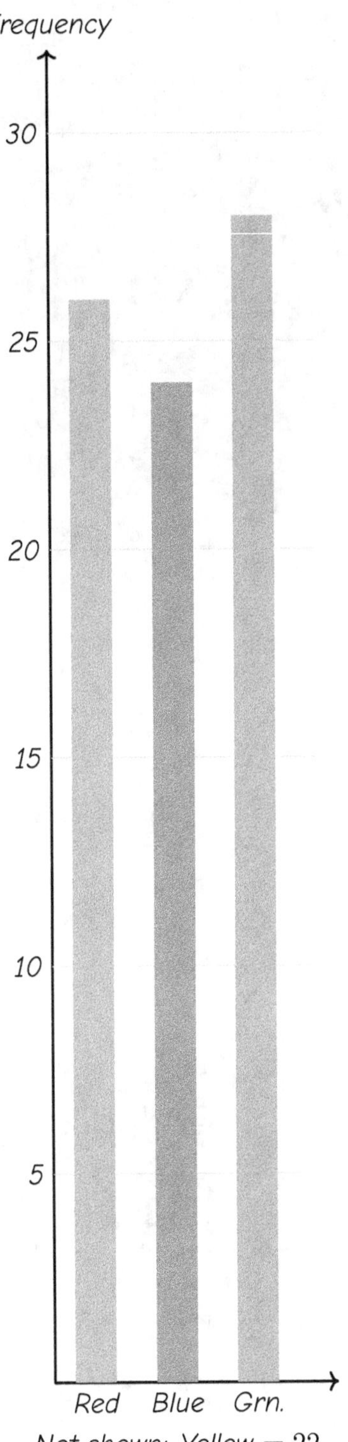

Not shown: Yellow = 22

(A) No, because the bars are different heights

(B) Yes, the frequencies $(26, 24, 28, 22)$ are all roughly close to 25

(C) Green is clearly the most likely color

(D) Cannot be determined from the graph

 # End of Practice Test 2 

Great job finishing the test!

 My Score

I got __________ out of 30 questions right.

*Check your answers in the **Answer Key** at the back of the book.*

💡 *Review any questions you missed. That's how we learn!*

📊 Check Your Score Online!

Visit **ViewMath Academy** to enter your answers and see which topics you need to review. You can also explore lessons, take quizzes, track your scores, and save your progress!

viewmath.com/score/7.1.IN.17

Or go to viewmath.com/score and enter code: 7.1.IN.17

Practice Test 3

 30 Questions

- ✔ **Read each question carefully** before choosing your answer.
- ✔ **Show your work** on scratch paper when you need to.
- ✔ **Skip hard questions** and come back to them later.
- ✔ **Check your answers** when you're done.
- ✔ **Take your time** — there's no rush!

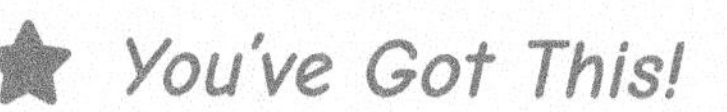

Do your best and show what you know!

1. *A proportional table has $k = \frac{3}{4}$. Which pair could be in the table?*

   (A) $(8, 10)$

   (B) $(8, 6)$

   (C) $(12, 8)$

   (D) $(4, 2)$

2. Two proportional lines are graphed. Line A is steeper than Line B. What does this tell you?

   (A) Line A has a smaller unit rate

   (B) Line A has a greater unit rate

   (C) Line B covers more distance

   (D) Both lines have the same rate

3. Store A sells 3 notebooks for $7.50. Store B sells 5 notebooks for $11.25. Which store offers the better deal?

   (A) Store A at $2.50 each

   (B) Store B at $2.25 each

   (C) Store A at $2.25 each

   (D) They cost the same

4. Sarah and Jake both solve "What is 40% of 75?" Sarah uses a proportion; Jake uses an equation. Which statement is true?

   (A) Only the proportion method works

   (B) Only the equation method works

   (C) Both methods give the same answer

   (D) The methods give different answers

5. Estimate a 15% tip on a $78 bill using the 10% method.

   (A) About $7.80

   (B) About $10.40

   (C) About $11.70

   (D) About $15.60

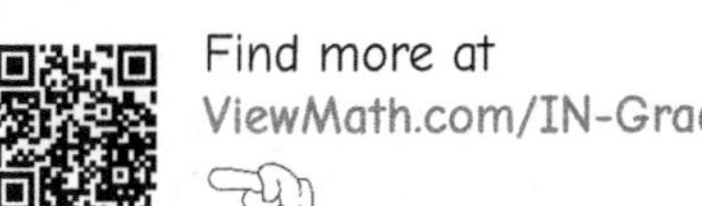
Find more at
ViewMath.com/IN-Grade7

6. A student measured a table as 4.5 feet long. The actual length is 5 feet. What is the percent error?

(A) 0.5%

(B) 5%

(C) 10%

(D) 11.1%

7. For which value of n is $|n| = 10$?

(A) Only $n = 10$

(B) Only $n = -10$

(C) $n = 10$ or $n = -10$

(D) $n = 0$

8. What is $15 + (-15)$?

(A) 30

(B) -30

(C) 15

(D) 0

9. Simplify $\frac{3}{4}n - \frac{1}{4}n + 5$.

(A) $\frac{1}{2}n + 5$

(B) $n + 5$

(C) $\frac{3}{4}n + 5$

(D) $\frac{2}{4}n - 5$

Find more at
ViewMath.com/IN-Grade7

10. Expand $\frac{2}{3}(9x + 6)$.

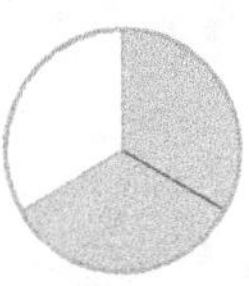

(A) $6x + 6$ (B) $6x + 4$

(C) $3x + 4$ (D) $18x + 4$

11. Solve $-1.5x + 4.5 = -3$.

(A) $x = -5$ (B) $x = 1$

(C) $x = 5$ (D) $x = -1$

12. Solve $-7x - 4 < 17$.

Your Answer:

13. Solve $-2x \leq 10$ and describe the graph.

(A) Closed circle at -5, shade left   (B) Closed circle at 5, shade left

(C) Closed circle at -5, shade right   (D) Open circle at -5, shade right

14. A blueprint uses 1 in = 9 ft. A room is 45 ft long in real life. How long is the room on the blueprint?

Your Answer:

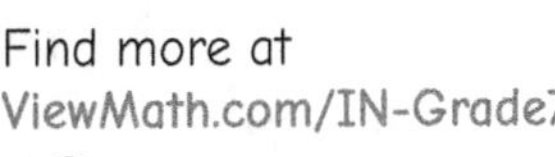

Find more at
ViewMath.com/IN-Grade7

15. Marcus wants to draw a triangle with angles 60°, 60°, and 60° and a side of 4 cm. How many different triangles can he draw?

(A) None

(B) Exactly one

(C) Exactly three

(D) Infinitely many

16. Two angles of a triangle are 25° and 25°. The triangle is:

(A) Equilateral

(B) Isosceles

(C) Scalene

(D) Impossible

17. The diagram shows a cone being sliced. Which cut produces a circular cross-section?

(A) Cut A (horizontal)

(B) Cut B (vertical through apex)

(C) Both cuts

(D) Neither cut

18. An angle is 4 times its supplement. What is the angle?

(A) 36°

(B) 45°

(C) 120°

(D) 144°

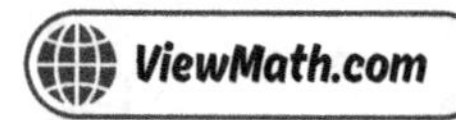

19. Point $M(3, -5)$ is reflected over the x-axis and then over the y-axis. What are the final coordinates?

(A) $(3, 5)$

(B) $(-3, -5)$

(C) $(-3, 5)$

(D) $(3, -5)$

20. A circular rug has a diameter of 8 feet. What is the area of the rug? Use $\pi \approx 3.14$

(A) $25.12 \ ft^2$

(B) $50.24 \ ft^2$

(C) $200.96 \ ft^2$

(D) $64 \ ft^2$

21. A figure is made of a right triangle with legs 6 cm and 8 cm, and a rectangle 8 cm by 3 cm. What is the total area?

(A) $24 \ cm^2$

(B) $48 \ cm^2$

(C) $36 \ cm^2$

(D) $72 \ cm^2$

22. The net shown below folds into a rectangular prism. Find the total surface area.

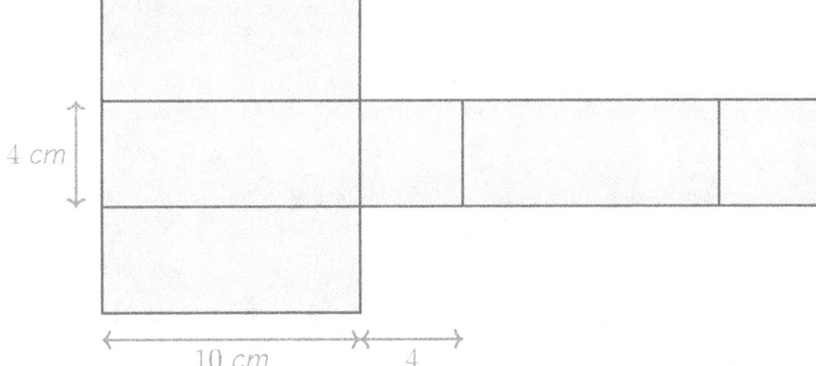

Your Answer

23. A trapezoidal prism has a trapezoidal base with parallel sides 4 cm and 6 cm, height 3 cm, and the prism length is 10 cm. What is its volume?

(A) $60 \ cm^3$

(B) $120 \ cm^3$

(C) $150 \ cm^3$

(D) $180 \ cm^3$

24. Which of the following increases the reliability of conclusions drawn from a sample?

(A) Using a smaller sample

(B) Using a larger random sample

(C) Surveying only people who agree with you

(D) Using a non-random sample

25. School A: median score 88, MAD 4. School B: median score 82, MAD 4. The difference in medians expressed in MADs is:

(A) 1 MAD

(B) 1.5 MADs

(C) 2 MADs

(D) 4 MADs

26. Group X: mean 45, MAD 6. Group Y: mean 57, MAD 6. Express the difference in means as a multiple of MAD.

Your Answer

27. Which statement is true about frequency tables?

(A) They only work for numerical data

(B) They can organize both categorical and numerical data

(C) They always require intervals

(D) They show data as a graph

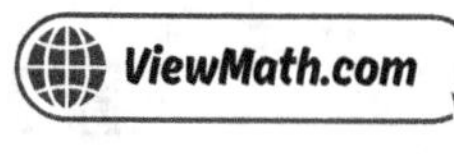

28. A bag contains 4 red marbles and 1 blue marble. Which type of probability model does this represent?

(A) Uniform

(B) Non-uniform

(C) Both uniform and non-uniform

(D) Neither

29. Two dice are rolled. What is the probability of getting a sum of 2?

(A) $\frac{1}{6}$

(B) $\frac{1}{12}$

(C) $\frac{1}{36}$

(D) $\frac{2}{36}$

30. You simulate flipping a coin 3 times and counting how many times all 3 are heads. In 40 simulations, all 3 heads occurs 6 times. What is the experimental probability?

(A) $\frac{6}{40} = 0.15$

(B) $\frac{1}{8} = 0.125$

(C) $\frac{3}{40} = 0.075$

(D) $\frac{6}{120} = 0.05$

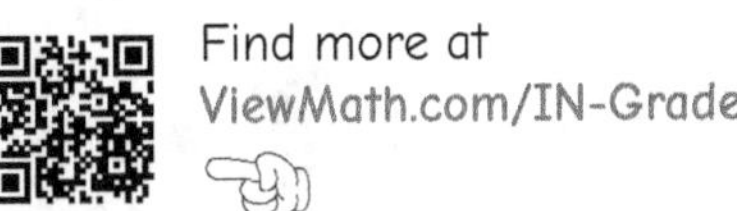

 *End of Practice Test 3* 

Great job finishing the test!

 My Score

I got _____________ out of 30 questions right.

*Check your answers in the **Answer Key** at the back of the book.*

Review any questions you missed. That's how we learn!

Check Your Score Online!

Visit **ViewMath Academy** to enter your answers and see which topics you need to review. You can also explore lessons, take quizzes, track your scores, and save your progress!

viewmath.com/score/7.1.IN.18

Or go to viewmath.com/score and enter code: 7.1.IN.18

4

Practice Test 4

☑ *30 Questions*

✏ Before You Start ✏

- ✔ **Read each question carefully** before choosing your answer.
- ✔ **Show your work** on scratch paper when you need to.
- ✔ **Skip hard questions** and come back to them later.
- ✔ **Check your answers** when you're done.
- ✔ **Take your time** — there's no rush!

⭐ You've Got This! ⭐

Do your best and show what you know!

1. What is the constant of proportionality (k) for the table below?

x	y
3	12
5	20
8	32

(A) 3

(B) 4

(C) 8

(D) 12

2. A proportional graph shows total cost versus hours of tutoring. The point $(3, 67.50)$ is on the graph. How much does 5 hours of tutoring cost?

Your Answer:

3. A factory produces 168 items in 7 hours. Another factory produces 200 items in 8 hours. Which factory is more productive?

(A) Factory 1, at 24 items/hr

(B) Factory 2, at 25 items/hr

(C) They produce the same rate

(D) Factory 1, at 25 items/hr

4. A school garden has 80 plants. If 45% are tomato plants, how many tomato plants are there?

Your Answer:

5. A restaurant bill is $40. You leave a 15% tip. How much is the tip?

(A) $4.00

(B) $5.00

(C) $6.00

(D) $8.00

Find more at
ViewMath.com/IN-Grade7

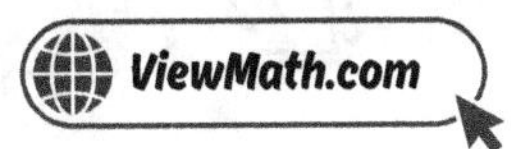

6. In the percent error formula, why do we use absolute value?

   (A) To make the calculation easier

   (B) Because the answer must be positive

   (C) Because we only care about the size of the error, not the direction

   (D) To convert the answer to a percent

7. Which statement about absolute value is FALSE?

   (A) $|-5| = 5$

   (B) $|0| = 0$

   (C) Absolute value can be negative

   (D) $|3| = |-3|$

8. A hiker starts at an elevation of -30 feet (below sea level). She climbs 45 feet. What is her new elevation?

   Your Answer:

9. How many terms are in the simplified form of $8b - 3 + 2b + 5 - b$?

   (A) 5

   (B) 3

   (C) 2

   (D) 1

10. Expand $5(2a - 3)$.

    (A) $10a - 3$

    (B) $7a - 3$

    (C) $10a - 15$

    (D) $10a + 15$

11. Solve $\dfrac{n}{2} + \dfrac{n}{3} = 10$.

    (A) $n = 15$

    (B) $n = 12$

    (C) $n = 6$

    (D) $n = 20$

Find more at
ViewMath.com/IN-Grade7

ViewMath.com

12. Write an inequality: "A number x divided by 4, minus 3, is at least 2." Then solve.

Your Answer:

13. Which graph correctly shows the solution to $2x - 5 > 3$?

Graph A:

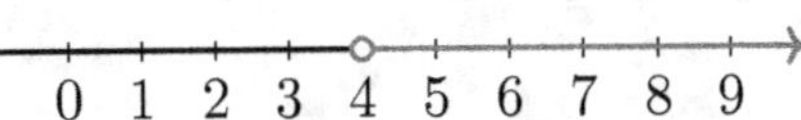

Graph B:

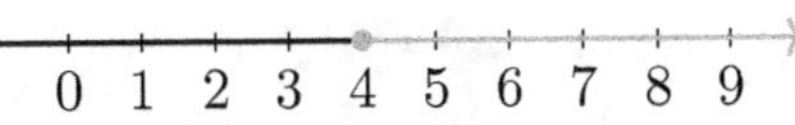

Graph C:

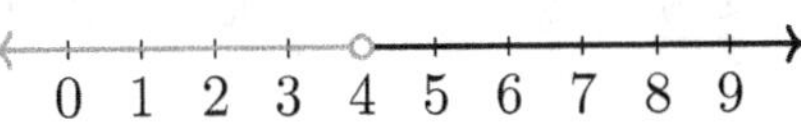

(A) Graph A

(B) Graph B

(C) Graph C

(D) None of the above

14. The actual length of a bridge is 360 m. On a scale drawing, the bridge is 12 cm. What is the scale in the form "1 cm = __ m"?

Your Answer:

15. You know all three angles of a triangle: 50°, 60°, and 70°. How many triangles can be drawn?

(A) None

(B) Exactly one

(C) Exactly three

(D) More than one (infinitely many)

16. *The diagram shows two triangles. Both have angles 30°, 60°, and 90°, but different side lengths. What does this demonstrate?*

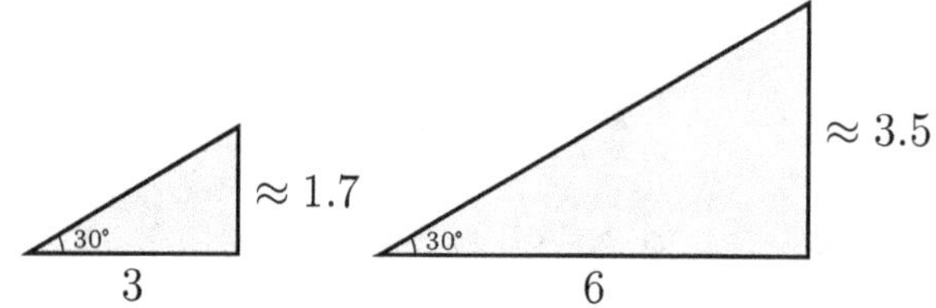

(A) SSS produces multiple triangles

(B) AAA produces a unique triangle

(C) AAA produces multiple triangles of different sizes

(D) These two triangles are congruent

17. *Which cross-section is NOT possible from slicing a rectangular prism?*

(A) Rectangle

(B) Triangle

(C) Circle

(D) Pentagon

18. *Two lines intersect. One angle formed is 130°. What are the measures of all four angles?*

(A) 130°, 130°, 50°, 50°

(B) 130°, 50°, 130°, 100°

(C) 130°, 130°, 130°, 130°

(D) 130°, 60°, 130°, 40°

19. *A rectangle has vertices $A(1,1)$, $B(5,1)$, $C(5,3)$, $D(1,3)$. It is translated left 3 and up 4. What are the coordinates of C'?*

(A) $(2,7)$

(B) $(8,7)$

(C) $(2,-1)$

(D) $(8,-1)$

20. What is the area of a circle with a radius of 10 cm? Use $\pi \approx 3.14$

   (A) $31.4 \ cm^2$ (B) $62.8 \ cm^2$

   (C) $100 \ cm^2$ (D) $314 \ cm^2$

21. A shape is made of a square with side 8 cm and a semicircle with diameter 8 cm attached to one side. What is the area? Use $\pi \approx 3.14$

   (A) $64 \ cm^2$ (B) $89.12 \ cm^2$

   (C) $114.24 \ cm^2$ (D) $139.12 \ cm^2$

22. A rectangular prism has dimensions 7 m by 4 m by 3 m. What is the area of the largest face?

   (A) $12 \ m^2$ (B) $21 \ m^2$

   (C) $28 \ m^2$ (D) $84 \ m^2$

23. The figure shows unit cubes stacked into an L-shape. Each cube has a side length of 1 cm. What is the total volume?

   (A) $8 \ cm^3$ (B) $11 \ cm^3$

   (C) $15 \ cm^3$ (D) $33 \ cm^3$

24. Give one reason why a sample might not perfectly represent a population.

Your Answer:

25. When comparing two populations visually, you should look at:

(A) Only the center (mean or median)

(B) Only the spread (range or MAD)

(C) Both the center and the spread

(D) Only the largest and smallest values

26. Store A daily sales: mean $500, MAD $40. Store B daily sales: mean $620, MAD $40. The difference in means in MADs is:

(A) 1.5 MADs

(B) 3 MADs

(C) 40 MADs

(D) 120 MADs

27. A line graph shows a plant's height each week: Week 1: 2 cm, Week 2: 5 cm, Week 3: 9 cm, Week 4: 14 cm. Between which weeks did the plant grow the most?

(A) Week 1 to Week 2

(B) Week 2 to Week 3

(C) Week 3 to Week 4

(D) Growth was the same each week

28. A coin is weighted so that heads comes up 60% of the time. What is the probability of tails?

(A) 0.60

(B) 0.50

(C) 0.40

(D) 0.30

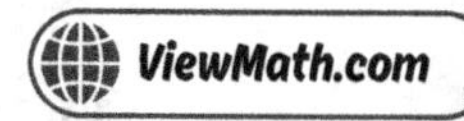

29. Look at the tree diagram for flipping three coins. What is the probability of getting exactly 2 tails? Write your answer as a fraction in simplest form.

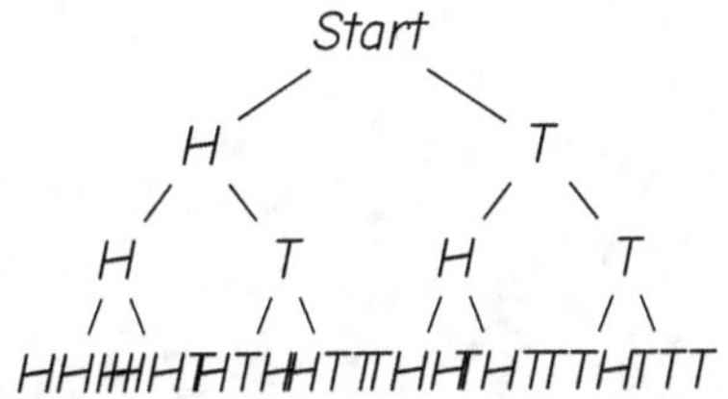

Your Answer:

30. A student wants to simulate the probability that exactly 2 out of 3 students pass a test, where each student has a 50% chance. Which model would work?

(A) Roll one die: odd = pass, even = fail

(B) Flip 3 coins: heads = pass, tails = fail; count exactly 2 heads

(C) Draw 3 marbles from a bag of 2 red and 1 blue

(D) Flip 1 coin 3 times and count all tails

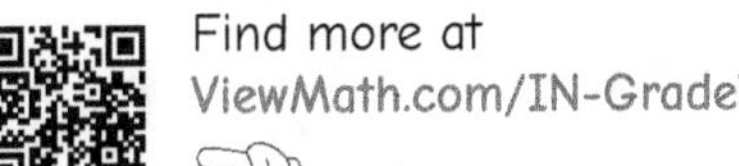
Find more at
ViewMath.com/IN-Grade7

 End of Practice Test 4 

Great job finishing the test!

 My Score

I got _____________ out of 30 questions right.

*Check your answers in the **Answer Key** at the back of the book.*

💡 *Review any questions you missed. That's how we learn!*

📊 Check Your Score Online!

Visit **ViewMath Academy** to enter your answers and see which topics you need to review. You can also explore lessons, take quizzes, track your scores, and save your progress!

viewmath.com/score/7.1.IN.19

Or go to viewmath.com/score and enter code: 7.1.IN.19

Practice Test 5

30 Questions

✏️ Before You Start ✏️

- ✔ **Read each question carefully** before choosing your answer.
- ✔ **Show your work** on scratch paper when you need to.
- ✔ **Skip hard questions** and come back to them later.
- ✔ **Check your answers** when you're done.
- ✔ **Take your time** — there's no rush!

⭐ You've Got This! ⭐

Do your best and show what you know!

1. A factory produces widgets at a constant rate. In $2\frac{1}{2}$ hours it makes 175 widgets. What is k in widgets per hour?

Your Answer:

2. A proportional graph passes through $(3, 18)$. What is the unit rate?

(A) 3

(B) 6

(C) 15

(D) 18

3. A 12-oz box of cereal costs \$3.60. A 20-oz box costs \$5.40. Which is the better value?

(A) 12-oz box at \$0.30/oz

(B) 20-oz box at \$0.27/oz

(C) 12-oz box at \$0.27/oz

(D) They are the same price per ounce

4. Look at the proportion model below. What value of n makes the proportion true?

$$\frac{n \quad\quad 80}{37.5 \quad\quad 100}$$

(A) 28

(B) 30

(C) 32

(D) 37.5

5. A real estate agent earns 3% commission on a \$250,000 home sale. What does she earn?

(A) \$750

(B) \$2,500

(C) \$7,500

(D) \$75,000

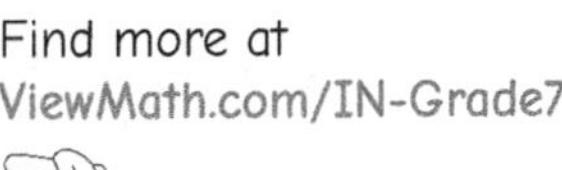
Find more at
ViewMath.com/IN-Grade7

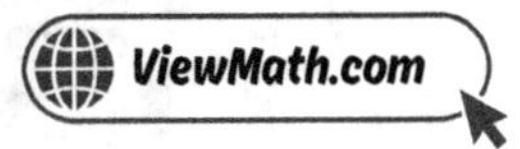

6. An architect estimated a building would be 45 meters tall. The actual height is 42 meters. What is the percent error (rounded to the nearest tenth)?

Your Answer:

7. Which statement is true?

   (A) $-5 > -2$

   (B) $-3 > 0$

   (C) $-7 < -4$

   (D) $0 < -1$

8. What value of n makes $(-9) + n = -2$ true?

   (A) -11

   (B) -7

   (C) 7

   (D) 11

9. Simplify $10 - 3x + 5x - 4$

   (A) $8x + 6$

   (B) $-8x + 6$

   (C) $2x + 6$

   (D) $2x - 6$

10. Expand and simplify $5(n - 3) - 2(n - 3)$.

Your Answer:

11. Solve $\dfrac{3x}{4} - 2 = 7$.

Your Answer:

Find more at
ViewMath.com/IN-Grade7

ViewMath.com

12. Solve $5x - 10 \geq 25$.

(A) $x \geq 3$

(B) $x \geq 7$

(C) $x \leq 7$

(D) $x \geq 35$

13. Solve $10 - 4x < 2$. Describe the graph.

Your Answer:

14. A map has a scale of $1\ cm = 50\ km$. Keisha measures the distance between two cities as $3.2\ cm$. What is the actual distance?

(A) 150 km

(B) 160 km

(C) 320 km

(D) 15.6 km

15. A rectangle must have a perimeter of 24 cm. How many different rectangles can satisfy this condition?

(A) None

(B) Exactly one

(C) Exactly two

(D) More than one

16. What condition (SSS, SAS, ASA, AAS, or AAA) gives infinitely many triangles?

Your Answer:

17. Which statement about cross-sections is true?

(A) The cross-section of any 3D figure is always a circle

(B) The cross-section depends on the angle and position of the cut

(C) A cross-section is always the same shape as the base

(D) Only prisms can have cross-sections

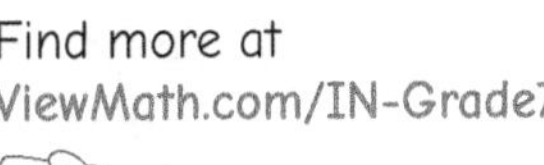
Find more at
ViewMath.com/IN-Grade7

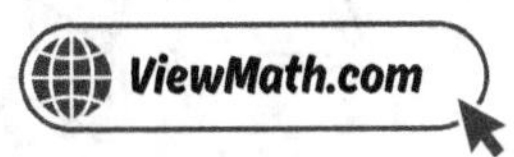

18. Two vertical angles are formed by intersecting lines. One angle is 72°. What is the vertical angle?

(A) 18°

(B) 72°

(C) 108°

(D) 288°

19. Which of the following is true about all three rigid transformations (translation, reflection, rotation)?

(A) They all change the size of the figure

(B) They all preserve the size and shape of the figure

(C) They all keep the figure's orientation the same

(D) They all move the figure to the right

20. A sprinkler waters a circular area with a radius of 5 meters. How many square meters does it water? Use $\pi \approx 3.14$.

(A) $31.4\ m^2$

(B) $50\ m^2$

(C) $78.5\ m^2$

(D) $157\ m^2$

21. An H-shaped figure is made of three rectangles. The top and bottom bars are each 10 cm by 2 cm. The middle vertical bar is 2 cm by 6 cm. What is the total area?

Your Answer:

22. How many faces does a rectangular prism have?

(A) 4

(B) 5

(C) 6

(D) 8

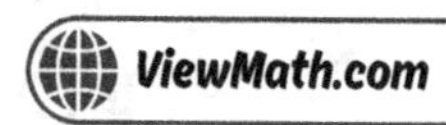

23. A moving box is 2 ft long, 2 ft wide, and 3 ft tall. How many of these boxes can fit in a space that is 6 ft by 4 ft by 3 ft?

(A) 3

(B) 6

(C) 12

(D) 18

24. A school has 800 students. The principal surveys 50 students about their favorite subject. What is the population?

(A) The 50 surveyed students

(B) All 800 students in the school

(C) The principal

(D) The students who like math

25. Team A's scores: mean 72, MAD 3. Team B's scores: mean 74, MAD 8. Which team is more consistent?

(A) Team A because its MAD is smaller

(B) Team B because its mean is higher

(C) Team A because its mean is lower

(D) Team B because its MAD is larger

26. Two groups have the same mean but different MADs. What does this tell you?

(A) One group scored higher than the other

(B) The groups have different levels of variability

(C) The groups are identical

(D) The means were calculated incorrectly

27. A line graph shows monthly rainfall: Jan 2 in., Feb 3 in., Mar 5 in., Apr 4 in. Find the total rainfall and the month with the most.

Your Answer:

28. A bag has tiles labeled A, A, A, B, B. What is the probability of drawing a B?

(A) $\frac{1}{5}$

(B) $\frac{2}{3}$

(C) $\frac{2}{5}$

(D) $\frac{3}{5}$

29. The table shows all possible products when two dice are rolled. How many outcomes give a product of 6? Write the probability as a fraction in simplest form.

×	1	2	3	4	5	6
1	1	2	3	4	5	6
2	2	4	6	8	10	12
3	3	6	9	12	15	18
4	4	8	12	16	20	24
5	5	10	15	20	25	30
6	6	12	18	24	30	36

Your Answer:

30. Which of the following is the FIRST step in designing a simulation?

(A) Record the results

(B) Run many trials

(C) Identify the event and its possible outcomes

(D) Calculate the experimental probability

Find more at
ViewMath.com/IN-Grade7

 # End of Practice Test 5 

Great job finishing the test!

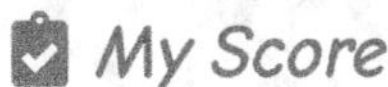 My Score

I got ___________ out of 30 questions right.

*Check your answers in the **Answer Key** at the back of the book.*

💡 *Review any questions you missed. That's how we learn!*

📊 Check Your Score Online!

Visit **ViewMath Academy** to enter your answers and see which topics you need to review. You can also explore lessons, take quizzes, track your scores, and save your progress!

viewmath.com/score/7.1.IN.20

Or go to viewmath.com/score and enter code: 7.1.IN.20

6

Practice Test 6

 30 Questions

✏ Before You Start ✏

- ✔ **Read each question carefully** before choosing your answer.
- ✔ **Show your work** on scratch paper when you need to.
- ✔ **Skip hard questions** and come back to them later.
- ✔ **Check your answers** when you're done.
- ✔ **Take your time** — there's no rush!

 ★ You've Got This! ★

Do your best and show what you know!

1. A proportional graph passes through $(6, 15)$. What is k?

Your Answer:

2. A proportional graph passes through $(4, 10)$. What is y when $x = 7$?

- (A) 13
- (B) 17
- (C) 17.5
- (D) 28

3. A car uses 3 gallons of gas every 87 miles. How many gallons does it need for a 435-mile trip?

- (A) 10
- (B) 12
- (C) 15
- (D) 18

4. 24 out of 80 students chose pizza for lunch. What percent is that?

- (A) 24%
- (B) 28%
- (C) 30%
- (D) 32%

5. An event planner charges a 10% planning fee plus a 3% coordination fee. If an event costs $4,000, what are the total fees?

- (A) $120
- (B) $400
- (C) $520
- (D) $1,300

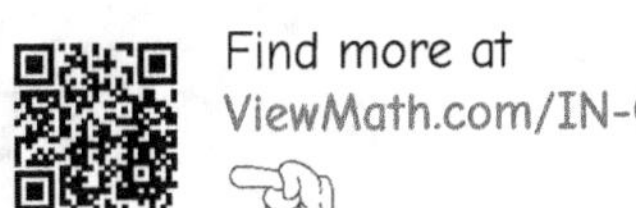
Find more at
ViewMath.com/IN-Grade7

6. A scientist predicted a chemical reaction would produce 30 grams. It actually produced 28 grams. What is the percent error?

(A) 2%

(B) 6.7%

(C) 7.1%

(D) 14.3%

7. What is the opposite of the opposite of 4?

(A) -4

(B) 0

(C) 4

(D) $\frac{1}{4}$

8. Which pair of integers has a sum of -6?

(A) 4 and -10

(B) -3 and -9

(C) 8 and -2

(D) -1 and -4

9. Simplify $9a - 4a + 6$.

(A) $11a$

(B) $5a + 6$

(C) $5a - 6$

(D) $13a - 4$

10. Expand $\frac{1}{2}(8x - 6)$.

(A) $4x - 6$

(B) $8x - 3$

(C) $4x - 3$

(D) $4x + 3$

11. Solve $-0.2n + 5.6 = 4$

   (A) $n = 8$ (B) $n = -8$

   (C) $n = 48$ (D) $n = -48$

12. Which inequality is represented by the number line below?

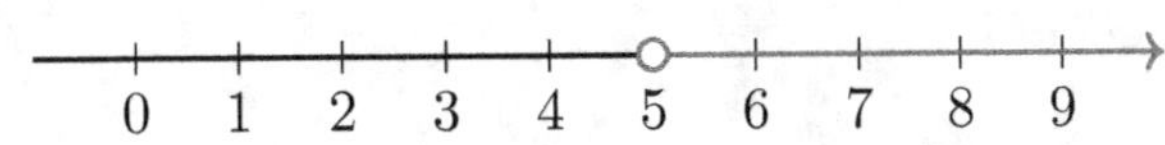

   (A) $x \geq 5$ (B) $x > 5$

   (C) $x \leq 5$ (D) $x < 5$

13. Solve $7x - 3 \geq 18$. Give the smallest integer that is a solution.

Your Answer:

14. A model car uses a scale of $1 : 24$. The real car is 180 inches long. How long is the model?

   (A) 7.5 inches   (B) 15 inches

   (C) 24 inches   (D) 156 inches

15. A square must have an area of 36 cm^2. How many different squares satisfy this condition?

   (A) None   (B) Exactly one

   (C) Exactly four   (D) More than one

16. Can a triangle have angles 90°, 100°, and −10°?

(A) Yes, it is a right triangle

(B) Yes, because the sum is 180°

(C) No, because an angle cannot be negative

(D) No, because the sum is not 180°

17. A rectangular prism has a base of 6 cm by 4 cm and a height of 10 cm. It is sliced parallel to the base. What are the dimensions of the cross-section?

Your Answer:

18. Three angles meet at a point on a line: 50°, $(2x + 10)$°, and x°. Find x.

Your Answer:

19. Point $E(-4, -1)$ is translated left 3 and up 5. What are the coordinates of E'?

(A) $(-7, 4)$

(B) $(-1, -6)$

(C) $(-1, 4)$

(D) $(-7, -6)$

20. A circle has an area of 12.56 m^2. Using $\pi \approx 3.14$, what is the diameter?

Your Answer:

21. A rectangular room is 14 ft by 10 ft. There is a built-in closet that is 3 ft by 4 ft. What is the area of the room not including the closet?

(A) 128 ft^2

(B) 140 ft^2

(C) 152 ft^2

(D) 12 ft^2

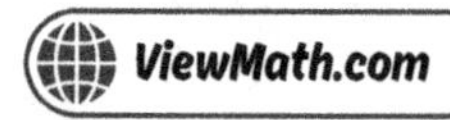

22. A cereal box is 25 cm tall, 18 cm wide, and 5 cm deep. How much cardboard is needed to make the box?

Your Answer:

23. A rectangular prism is 8 in by 5 in by 4 in. Another is 10 in by 4 in by 4 in. Which has more volume and by how much?

(A) First prism, by 10 in^3

(B) Second prism, by 10 in^3

(C) First prism, by 20 in^3

(D) They have the same volume

24. A random sample means that:

(A) Only the best members are chosen

(B) The largest possible group is chosen

(C) Every member of the population has an equal chance of being selected

(D) Only volunteers are selected

25. When is a dot plot a better choice than a box plot for comparing two groups?

(A) When data sets are very large

(B) When you want to see every individual data point

(C) When you only want to compare medians

(D) When the data is categorical

26. Data set: 12, 15, 18, 20, 22, 25, 28. What is the median?

(A) 18

(B) 20

(C) 22

(D) 19

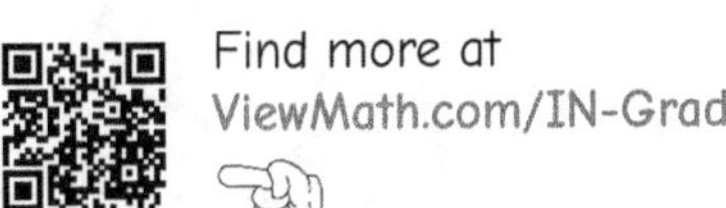
Find more at
ViewMath.com/IN-Grade7

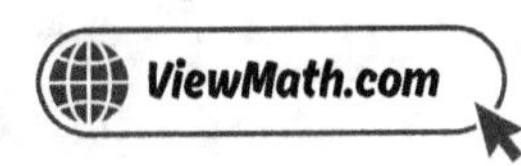

27. When would you choose a double bar graph instead of two separate bar graphs?

(A) When the data sets have different categories

(B) When you want to compare two data sets for the same categories

(C) When you want to show change over time

(D) When you want to show parts of a whole

28. A spinner is divided into 4 sections with probabilities: red $= 0.4$, blue $= 0.3$, green $= 0.2$, yellow $= 0.1$. What type of model is this?

(A) Uniform, because there are 4 sections

(B) Non-uniform, because the probabilities are different

(C) Invalid, because the probabilities don't add up to 1

(D) Uniform, because all sections are on the same spinner

29. A coin is flipped and a die is rolled. What is the probability of getting tails and an even number?

(A) $\frac{1}{12}$

(B) $\frac{1}{6}$

(C) $\frac{1}{4}$

(D) $\frac{1}{2}$

30. What is a simulation?

(A) A way to calculate exact probabilities

(B) A model that uses random numbers or objects to imitate a real event

(C) A method that always gives the same result

(D) A graph that shows all possible outcomes

 *End of Practice Test 6* 

Great job finishing the test!

 My Score

I got _____________ out of 30 questions right.

Check your answers in the **Answer Key** at the back of the book.

💡 Review any questions you missed. That's how we learn!

📊 Check Your Score Online!

Visit **ViewMath Academy** to enter your answers and see which topics you need to review. You can also explore lessons, take quizzes, track your scores, and save your progress!

viewmath.com/score/7.1.IN.21

Or go to viewmath.com/score and enter code: 7.1.IN.21

Practice Test 7

✅ 30 Questions

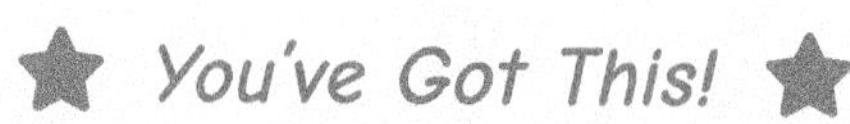

✏️ Before You Start ✏️

- ✔ **Read each question carefully** before choosing your answer.
- ✔ **Show your work** on scratch paper when you need to.
- ✔ **Skip hard questions** and come back to them later.
- ✔ **Check your answers** when you're done.
- ✔ **Take your time** — there's no rush!

⭐ You've Got This! ⭐

Do your best and show what you know!

1. Find the missing value in the proportional table.

x	y
5	20
9	?

Your Answer:

2. Which statement about proportional graphs is FALSE?

(A) They always pass through the origin

(B) The slope equals the unit rate

(C) A steeper line means a larger k

(D) The line can be curved

3. A cyclist rides 42 miles in 3 hours. At this rate, how long does it take to ride 98 miles?

Your Answer:

4. 135 is 54% of what number?

(A) 72.9

(B) 200

(C) 245

(D) 250

5. An online marketplace charges a 4% seller fee. A seller lists an item for $85. How much does the seller receive after the fee?

Your Answer:

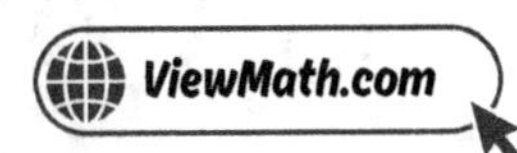

6. You estimated there were 50 marbles in a jar. The actual count was 40. What is the percent error?

(A) 10%

(B) 20%

(C) 25%

(D) 50%

7. Which integer is 5 units to the left of 2 on the number line?

Your Answer

8. A diver is at -20 feet. She rises 8 feet. What is her new depth?

(A) -28 feet

(B) -12 feet

(C) 12 feet

(D) 28 feet

9. Which pair of terms are like terms?

(A) $3x$ and $3y$

(B) $5m$ and $-2m$

(C) $4n$ and $4n^2$

(D) 7 and $7x$

10. Which expression is equivalent to $7(a - 3) - 2a$?

(A) $5a - 3$

(B) $5a + 21$

(C) $9a - 21$

(D) $5a - 21$

11. A store marks down a jacket by $\frac{1}{4}$ of its original price p, then subtracts an additional \$8. The sale price is \$37. Find the original price.

Your Answer

Find more at
ViewMath.com/IN-Grade7

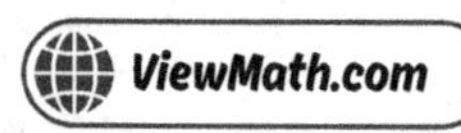

12. A movie theater requires you to be more than 48 inches tall for a ride. You are 42 inches tall and grow g inches per year. Which inequality finds how many years until you are tall enough?

(A) $42 + g > 48$

(B) $42g > 48$

(C) $48 - g > 42$

(D) $42 - g > 48$

13. Solve $4x + 3 \leq 19$ and describe the graph.

(A) Closed circle at 4, shade left

(B) Open circle at 4, shade left

(C) Closed circle at 4, shade right

(D) Closed circle at 5.5, shade left

14. Which statement about scale drawings is always true?

(A) Angles in the drawing are different from the actual angles

(B) The scale factor changes depending on which part of the drawing you measure

(C) All lengths in the drawing are multiplied by the same scale factor to get actual lengths

(D) If the scale factor is 4, the area of the actual figure is 4 times the drawing area

15. Can a triangle have sides 10 cm, 6 cm, and 3 cm? Explain why or why not.

Your Answer:

16. A triangle has angles 40°, 60°, and 80°. How many different triangles can be drawn?

(A) None

(B) Exactly one

(C) Exactly three

(D) Infinitely many

  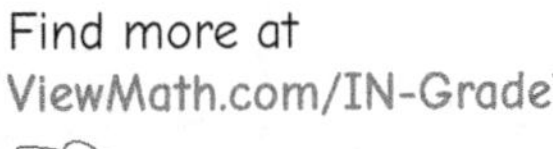 

Find more at
ViewMath.com/IN-Grade7

ViewMath.com

17. You slice a rectangular prism with a cut parallel to its base. What shape is the cross-section?

(A) Triangle

(B) Circle

(C) Rectangle

(D) Pentagon

18. Two supplementary angles are equal. What is the measure of each?

(A) 45°

(B) 60°

(C) 90°

(D) 180°

19. Point $G(1, 4)$ is rotated 90° clockwise about the origin. Then the result is reflected over the x-axis. What are the final coordinates?

(A) $(4, 1)$

(B) $(-4, -1)$

(C) $(4, -1)$

(D) $(-4, 1)$

20. Find the area of a semicircle with a diameter of 10 cm. Use $\pi \approx 3.14$.

Your Answer:

21. A T-shaped figure can be broken into a 12 cm by 2 cm horizontal rectangle and a 2 cm by 8 cm vertical rectangle. What is the total area?

(A) 24 cm^2

(B) 32 cm^2

(C) 40 cm^2

(D) 48 cm^2

22. What does surface area measure?

(A) The space inside a 3D figure

(B) The distance around a 3D figure

(C) The total area of all faces of a 3D figure

(D) The height of a 3D figure

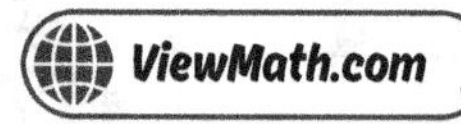

23. Find the volume of a rectangular prism with dimensions 9 cm by 5 cm by 4 cm.

Your Answer:

24. The diagram below shows four different sampling methods used to survey students at a school. Which method is most likely to produce an unbiased sample?

Method A
Survey the football team

Method B
Survey every 10th student
on the school roster

Method C
Post survey online and
wait for responses

Method D
Survey students in
one math class

(A) Method A

(B) Method B

(C) Method C

(D) Method D

25. Class A: median 90, IQR 6. Class B: median 83, IQR 14. Which class is more consistent?

Your Answer:

26. Class A: median 85, IQR 8, range 20. Class B: median 80, IQR 15, range 40. A student says "Class B is better because it has a wider range." Is the student correct?

(A) Yes — wider range means more talent

(B) No — wider range means less consistency, and Class A has a higher median

(C) Yes — Class B's scores spread more, which is always better

(D) No — range does not measure anything useful

27. The line graph shows the number of visitors to a library each day for one week.

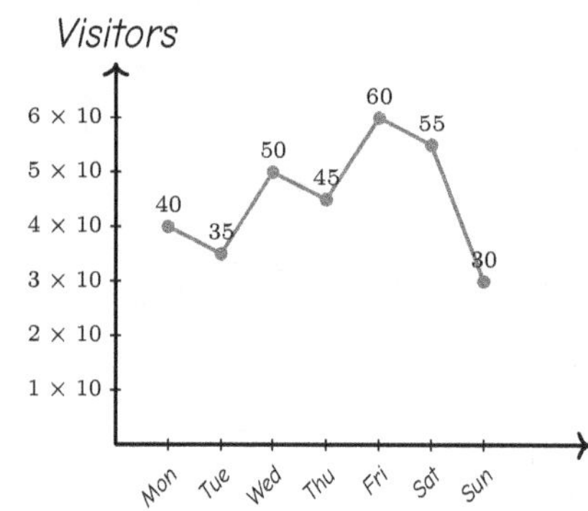

Find the mean number of daily visitors and the range. On which day was the library busiest?

Your Answer:

28. Which of the following is an example of a uniform probability model?

(A) A bag with 3 red, 5 blue, and 2 green marbles

(B) A fair number cube with 6 sides

(C) A spinner with sections of different sizes

(D) A weighted coin

29. A coin is flipped and a die is rolled. What is the probability of getting heads and a number greater than 4?

(A) $\frac{1}{12}$

(B) $\frac{1}{6}$

(C) $\frac{2}{12}$

(D) $\frac{1}{3}$

30. A game has a 25% chance of winning. Which simulation model correctly represents this?

(A) Flip a coin — heads means win

(B) Roll a die — roll a 1 means win

(C) Use a spinner with 4 equal sections — one section means win

(D) Use digits 0–9 — digits 0–4 mean win

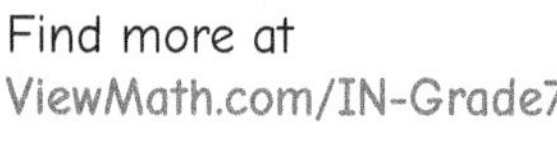

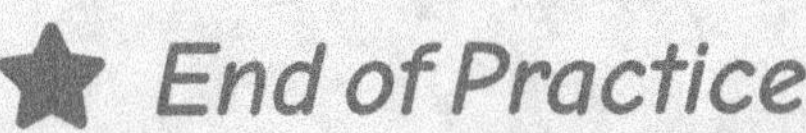 **End of Practice Test 7**

Great job finishing the test!

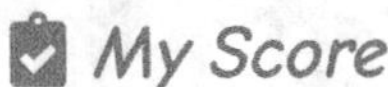 **My Score**

I got ___________ out of 30 questions right.

*Check your answers in the **Answer Key** at the back of the book.*

💡 *Review any questions you missed. That's how we learn!*

📊 **Check Your Score Online!**

Visit **ViewMath Academy** to enter your answers and see which topics you need to review. You can also explore lessons, take quizzes, track your scores, and save your progress!

viewmath.com/score/7.1.IN.22

Or go to viewmath.com/score and enter code: 7.1.IN.22

8

Practice Test 8

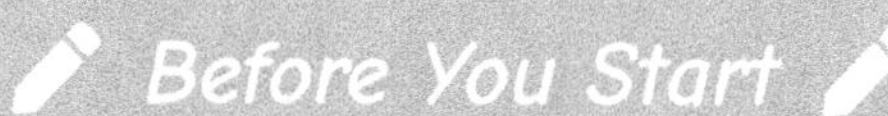
30 Questions

✏ Before You Start ✏

- ✔ **Read each question carefully** before choosing your answer.
- ✔ **Show your work** on scratch paper when you need to.
- ✔ **Skip hard questions** and come back to them later.
- ✔ **Check your answers** when you're done.
- ✔ **Take your time** — there's no rush!

⭐ You've Got This! ⭐

Do your best and show what you know!

1. A car travels at a constant speed. It goes 150 miles in 2.5 hours. What is the constant of proportionality?

(A) 30 mph

(B) 60 mph

(C) 75 mph

(D) 150 mph

2. Two friends start walking at the same time. The graph shows their distances. How many miles apart are they after 6 hours?

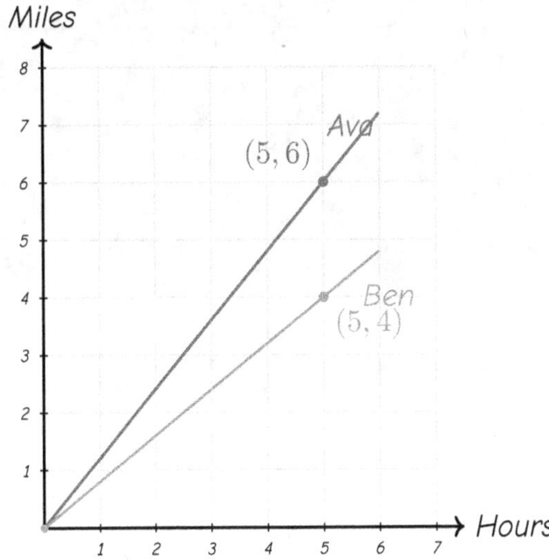

Your Answer:

3. A scale drawing uses 1 cm = 4.5 m. If a building is 27 m tall, how tall is it in the drawing?

Your Answer:

4. 27 out of 180 students earned an A on the exam. What percent earned an A?

(A) 12%

(B) 15%

(C) 18%

(D) 27%

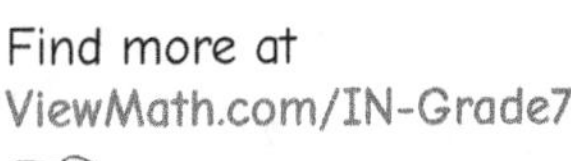 Find more at
ViewMath.com/IN-Grade7

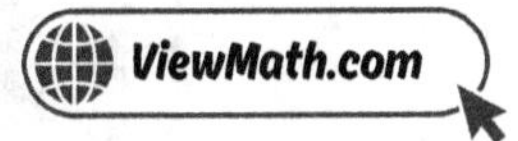 ViewMath.com

5. Estimate a 20% tip on a $93 bill. Round the bill first, then calculate.

Your Answer:

6. The number line below shows an estimated value and an actual value. What is the percent error?

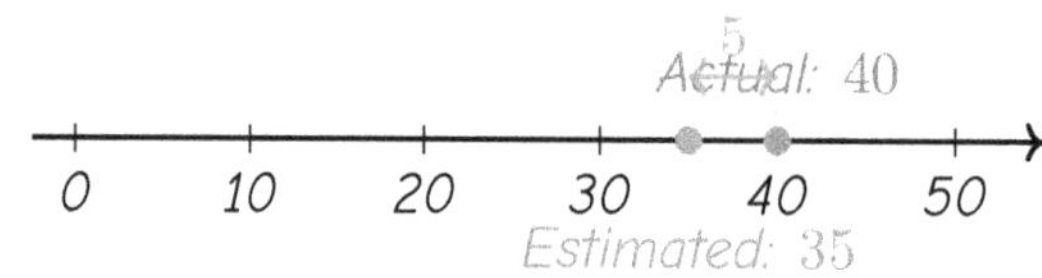

(A) 5%

(B) 10%

(C) 12.5%

(D) 14.3%

7. An elevator goes 3 floors up from the lobby, then 7 floors down. Which integer describes its final position relative to the lobby?

(A) 4

(B) −4

(C) 10

(D) −10

8. A bank account has −$50 (overdrawn). A deposit of $75 is made. What is the new balance?

(A) −$125

(B) −$25

(C) $25

(D) $125

9. Simplify $\frac{1}{3}x + \frac{2}{3}x$.

(A) $\frac{2}{9}x$ (B) x

(C) $\frac{3}{6}x$ (D) $\frac{1}{3}x^2$

10. Expand and simplify $-4(a - 2) + 3(a + 5) - 7$.

Your Answer:

11. Use the fraction bar model to solve: $\frac{2}{3}x = 8$. What is x?

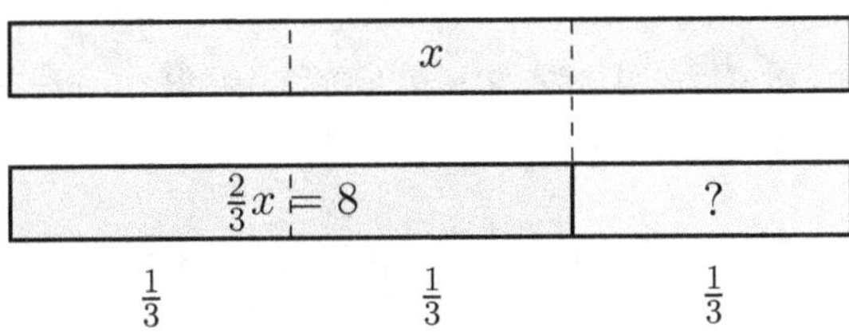

(A) $x = 12$ (B) $x = 16$

(C) $x = 10$ (D) $x = 5\frac{1}{3}$

12. Solve $3x + 7 \leq 25$.

(A) $x \leq \frac{32}{3}$ (B) $x \leq 6$

(C) $x \geq 6$ (D) $x \leq 9$

Find more at
ViewMath.com/IN-Grade7

ViewMath.com

13. What inequality is shown by a closed circle at -6 with shading to the right?

Your Answer:

14. A scale drawing of a soccer field is shown on the grid below. Each grid square represents $10\ m \times 10\ m$ in real life. What is the actual perimeter of the field?

5 squares

Field

9 squares

Your Answer:

15. Two sides of a triangle are 5 cm and 5 cm with an included angle of 60°. What type of triangle is formed?

Your Answer:

16. A triangle has angles 60°, 60°, and 60°, and one side measures 10 cm. How many triangles can be drawn?

Your Answer:

17. What shape is the cross-section when a cone is sliced vertically through its apex?

Your Answer:

Find more at
ViewMath.com/IN-Grade7

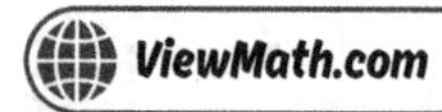

18. Which statement about vertical angles is always true?

(A) They are supplementary

(B) They are complementary

(C) They are equal

(D) They sum to 360°

19. A triangle has vertices $A(2, 1)$, $B(6, 1)$, and $C(6, 4)$. Reflect the triangle over the y-axis. List the new vertices.

Your Answer:

20. A square has a circle inscribed inside it as shown. The side of the square is 10 cm. What is the area of the circle? Use $\pi \approx 3.14$.

10 cm

(A) 31.4 cm^2

(B) 78.5 cm^2

(C) 100 cm^2

(D) 314 cm^2

21. A rectangular patio is 12 ft by 8 ft. A circular hot tub with a diameter of 6 ft is placed on the patio. What is the area of the patio NOT covered by the hot tub? Use $\pi \approx 3.14$.

Your Answer:

22. What is the surface area of a rectangular prism with length 5 cm, width 3 cm, and height 2 cm?

(A) $30\ cm^2$

(B) $62\ cm^2$

(C) $46\ cm^2$

(D) $31\ cm^2$

23. What is the volume of a cube with side length 5 m?

(A) $15\ m^3$

(B) $25\ m^3$

(C) $125\ m^3$

(D) $150\ m^3$

24. A random sample of 50 apples from an orchard found that 6 were bruised. If the orchard has 3,000 apples, about how many are bruised?

Your Answer:

25. Team A: mean 60, MAD 4. Team B: mean 70, MAD 3. Is the difference in means meaningful? Explain briefly.

Your Answer:

26. Data set: $20, 22, 25, 27, 30, 32, 35$. What is the IQR?

(A) 15

(B) 10

(C) 7

(D) 5

27. A line graph shows temperatures: Mon 72°, Tue 68°, Wed 70°, Thu 75°, Fri 73°. What is the range of temperatures?

Your Answer:

28. A bag contains 5 yellow and 5 purple marbles. Lisa picks a marble, records the color, and replaces it 40 times. She gets yellow 24 times. Which statement is correct?

(A) The model predicts $P(yellow) = 0.5$, and the experiment gives $P(yellow) = 0.6$, so there is a small difference

(B) The model is wrong because $24 \neq 20$

(C) The model predicts $P(yellow) = 0.6$

(D) The experiment proves yellow is more likely than purple

29. A coin is flipped and a die is rolled. What is the probability of getting tails and a 1? Write your answer as a fraction.

Your Answer:

30. How does increasing the number of trials in a simulation affect the results?

(A) The results become less reliable

(B) The experimental probability gets closer to the theoretical probability

(C) The results stay exactly the same

(D) The simulation takes less time

Find more at
ViewMath.com/IN-Grade7

 # End of Practice Test 8 

Great job finishing the test!

 My Score

I got ___________ out of 30 questions right.

*Check your answers in the **Answer Key** at the back of the book.*

💡 *Review any questions you missed. That's how we learn!*

📊 Check Your Score Online!

Visit **ViewMath Academy** to enter your answers and see which topics you need to review. You can also explore lessons, take quizzes, track your scores, and save your progress!

viewmath.com/score/7.1.IN.23

Or go to viewmath.com/score and enter code: 7.1.IN.23

Practice Test 9

📋 30 Questions

✏️ Before You Start ✏️

- ✔ **Read each question carefully** before choosing your answer.
- ✔ **Show your work** on scratch paper when you need to.
- ✔ **Skip hard questions** and come back to them later.
- ✔ **Check your answers** when you're done.
- ✔ **Take your time** — there's no rush!

⭐ You've Got This! ⭐

Do your best and show what you know!

1. Two students find k from the point $(8, 20)$. Student A says $k = 2.5$. Student B says $k = 0.4$. Who is correct?

   (A) Student A, because $k = \frac{y}{x}$

   (B) Student B, because $k = \frac{x}{y}$

   (C) Both are correct, depending on which variable is independent

   (D) Neither is correct

2. The equation $y = 5x$ gives the cost of movie tickets. What does the point $(1, 5)$ represent?

   (A) 5 movies cost $1

   (B) 1 ticket costs $5

   (C) The total for all tickets is $5

   (D) You need 5 hours to buy 1 ticket

3. A printer prints 120 pages in 8 minutes. How long will it take to print 450 pages?

   (A) 25 min

   (B) 28 min

   (C) 30 min

   (D) 36 min

4. In a bag of 150 marbles, 60 are blue. What percent of the marbles are blue?

   (A) 25%

   (B) 30%

   (C) 35%

   (D) 40%

5. A hair stylist charges $55 for a haircut. The customer leaves a 20% tip. What is the total the customer pays?

   (A) $11.00

   (B) $60.50

   (C) $66.00

   (D) $75.00

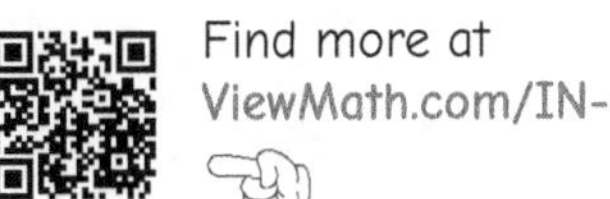

Find more at
ViewMath.com/IN-Grade7

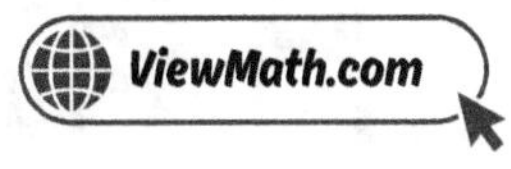

6. A builder estimated a project would cost \$8,000. The actual cost was \$7,500. What is the percent error?

   (A) 5%　　　　　　　　　　　　　　(B) 6.25%

   (C) 6.67%　　　　　　　　　　　　　(D) 10%

7. Which expression equals 0?

   (A) $7 + 7$　　　　　　　　　　　　(B) $7 - 7$

   (C) $7 \times 0 + 7$　　　　　　　　　(D) $|-7| - |7|$

8. A thermometer shows temperature changes over three hours.

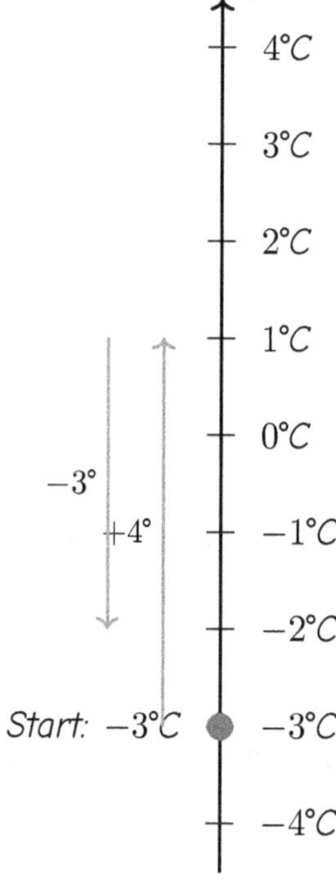

What is the final temperature after these two changes?

Your Answer:

Find more at
ViewMath.com/IN-Grade7

ViewMath.com

9. Which expression is equivalent to $6m + 4 - 2m + 3$?

(A) $4m + 7$

(B) $8m + 7$

(C) $4m + 1$

(D) $4m - 7$

10. Use the area model to expand $5(3x + 2)$. Fill in the missing values and write the expanded expression.

$$
\begin{array}{c|c|c}
 & 3x & 2 \\
\hline
5 & ? & ? \\
\end{array}
$$

Your Answer:

11. Solve $\dfrac{x}{6} - \dfrac{2}{3} = \dfrac{1}{2}$.

(A) $x = 3$

(B) $x = 7$

(C) $x = -1$

(D) $x = 10$

12. Solve $2x + 3 > 11$.

(A) $x > 4$

(B) $x > 7$

(C) $x < 4$

(D) $x > 14$

13. True or false: The graph of $x < 5$ and $x \le 5$ look exactly the same.

(A) True — both shade to the left of 5

(B) False — $x < 5$ has an open circle, $x \le 5$ has a closed circle

(C) False — they shade in different directions

(D) True — the circle type does not matter

Find more at
ViewMath.com/IN-Grade7

ViewMath.com

14. A blueprint uses the scale 1 in = 6 ft. A hallway is 3.5 in long on the blueprint. What is the actual length?

(A) 18 ft

(B) 21 ft

(C) 24 ft

(D) 9.5 ft

15. A parallelogram has side lengths 5 cm and 8 cm. How many different parallelograms can be drawn?

(A) None

(B) Exactly one

(C) Exactly two

(D) More than one

16. Can a triangle be formed with sides 7, 10, and 15?

(A) No, the triangle inequality fails

(B) Yes, exactly one triangle

(C) Yes, more than one triangle

(D) Yes, but only a right triangle

17. A triangular prism is sliced with a cut parallel to its triangular base. What is the cross-section?

(A) A rectangle

(B) A triangle congruent to the base

(C) A smaller triangle

(D) A trapezoid

18. Two supplementary angles are $(2x + 10)°$ and $(3x)°$. What is the value of x?

(A) 30

(B) 34

(C) 36

(D) 40

Find more at
ViewMath.com/IN-Grade7

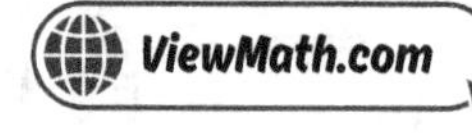

19. A triangle has vertices $P(1,2)$, $Q(4,2)$, and $R(4,6)$. After a translation, $P' = (3,0)$. What translation was applied?

(A) Right 2, down 2

(B) Left 2, up 2

(C) Right 2, up 2

(D) Left 2, down 2

20. A quarter-circle has a radius of 8 cm. What is its area? Use $\pi \approx 3.14$.

(A) $200.96\ cm^2$

(B) $100.48\ cm^2$

(C) $50.24\ cm^2$

(D) $25.12\ cm^2$

21. A pentagon can be divided into a rectangle 6 cm by 4 cm and a triangle with base 6 cm and height 2 cm. What is the total area?

(A) $24\ cm^2$

(B) $30\ cm^2$

(C) $36\ cm^2$

(D) $48\ cm^2$

22. What is the surface area of a cube with side length 4 cm?

(A) $16\ cm^2$

(B) $64\ cm^2$

(C) $96\ cm^2$

(D) $24\ cm^2$

23. A rectangular prism has a base area of $24\ cm^2$ and a height of 7 cm. What is its volume?

(A) $31\ cm^3$

(B) $168\ cm^3$

(C) $336\ cm^3$

(D) $17\ cm^3$

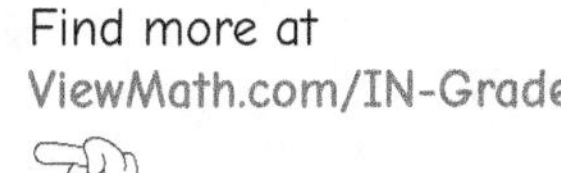

Find more at
ViewMath.com/IN-Grade7

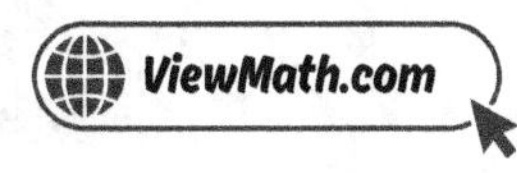

24. A researcher wants to know the average height of seventh graders in a state. She measures the heights of 100 seventh graders from different schools. What is the sample?

(A) All seventh graders in the state

(B) All students in the schools she visited

(C) The 100 seventh graders she measured

(D) The tallest seventh graders

25. The box plots below compare test scores for Class A and Class B.

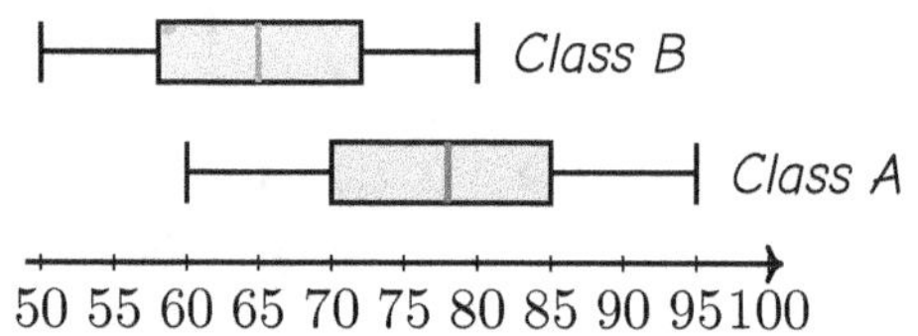

Which statement is best supported by the box plots?

(A) Class B scored higher than Class A

(B) Class A has a higher median and a larger IQR

(C) Class A has a higher median and both classes have the same IQR

(D) Class A and Class B overlap, but Class A's median is higher

26. Group A: $\{60, 65, 70, 75, 80\}$. Group B: $\{40, 55, 70, 85, 100\}$. Both have a mean of 70. Which group is more variable?

(A) Group A

(B) Group B

(C) They are equally variable

(D) Cannot be determined

Find more at
ViewMath.com/IN-Grade7

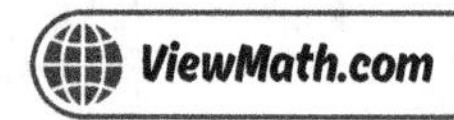

27. The double bar graph shows the number of goals scored by two soccer teams over four games.

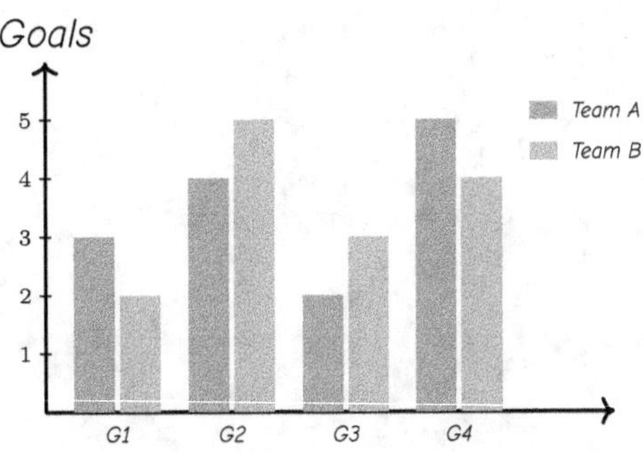

Which team scored more total goals?

(A) Team A with 14 goals

(B) Team B with 14 goals

(C) They tied with 14 goals each

(D) Team A with 12 goals

28. A teacher predicted that a spinner would land on red $\frac{1}{4}$ of the time. After 80 spins, red appeared 25 times. How do the observed and predicted results compare?

(A) They are very different — the model is wrong

(B) The observed ($\frac{25}{80} = 0.3125$) is close to the predicted (0.25)

(C) The observed result is exactly equal to the predicted result

(D) The predicted probability is higher than the observed

29. Three coins are flipped. What is the probability of getting NO heads (all tails)?

(A) $\frac{1}{2}$

(B) $\frac{1}{4}$

(C) $\frac{1}{8}$

(D) $\frac{3}{8}$

30. Describe how to use a standard die to simulate a $\frac{1}{3}$ probability event.

Your Answer:

 # End of Practice Test 9 

Great job finishing the test!

 My Score

I got _____________ out of 30 questions right.

*Check your answers in the **Answer Key** at the back of the book.*

 *Review any questions you missed. That's how we learn!*

Check Your Score Online!

Visit **ViewMath Academy** to enter your answers and see which topics you need to review. You can also explore lessons, take quizzes, track your scores, and save your progress!

viewmath.com/score/7.1.IN.24

Or go to viewmath.com/score and enter code: 7.1.IN.24

10

Practice Test 10

30 Questions

✏ Before You Start ✏

- ✔ **Read each question carefully** before choosing your answer.
- ✔ **Show your work** on scratch paper when you need to.
- ✔ **Skip hard questions** and come back to them later.
- ✔ **Check your answers** when you're done.
- ✔ **Take your time** — there's no rush!

⭐ You've Got This! ⭐

Do your best and show what you know!

1. The graph below shows a proportional relationship. What is the constant of proportionality?

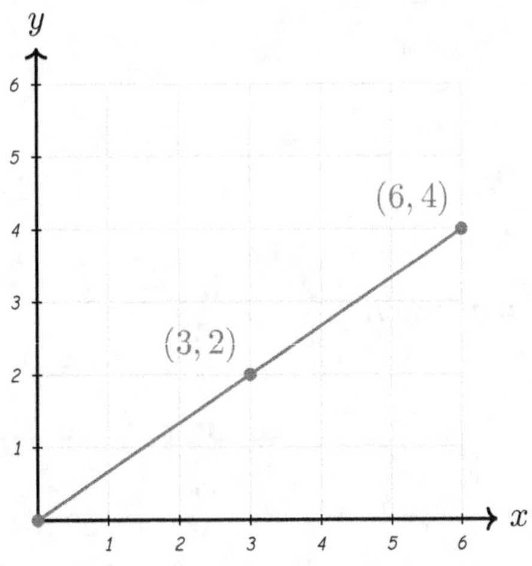

(A) $k = \frac{1}{2}$

(B) $k = \frac{2}{3}$

(C) $k = \frac{3}{2}$

(D) $k = 2$

2. The graph below shows a proportional relationship between hours and miles traveled. What is the unit rate?

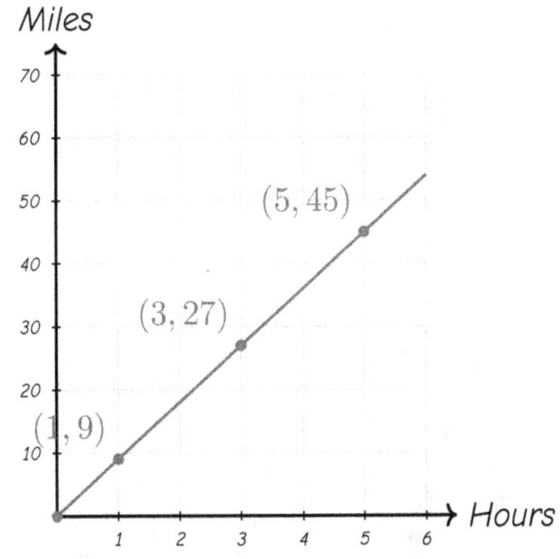

(A) 3 mph

(B) 9 mph

(C) 27 mph

(D) 45 mph

3. A map scale says $2 \text{ cm} = 15$ miles. Two cities are 7 cm apart on the map. What is the actual distance?

(A) 42.5 miles

(B) 45 miles

(C) 52.5 miles

(D) 105 miles

4. Which equation is equivalent to the proportion $\dfrac{n}{250} = \dfrac{16}{100}$?

(A) $n = 250 \div 16$

(B) $n = 16 \times 250$

(C) $100n = 250 \times 16$

(D) $16n = 250 \times 100$

5. Two friends split a \$70 dinner bill equally and each leave a 15% tip on their share. How much does each person pay in total?

(A) \$35.00

(B) \$37.25

(C) \$40.25

(D) \$45.50

6. The bar graph shows the estimated and actual amounts of water (in liters) collected during an experiment over three days. On which day was the percent error the largest, and what was it?

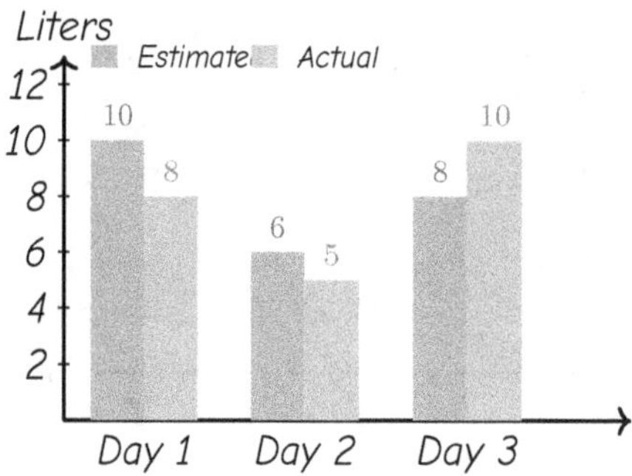

Your Answer:

7. Find $|-23|$.

Your Answer:

8. What is $(-18) + 18$?

Your Answer:

9. Simplify $3y + 8 - 5y - 2$.

(A) $8y + 6$

(B) $-2y + 6$

(C) $2y - 6$

(D) $-2y + 10$

10. Expand $-4(2y - 5)$.

(A) $-8y - 20$

(B) $-8y - 5$

(C) $-8y + 20$

(D) $8y + 20$

Find more at
ViewMath.com/IN-Grade7

ViewMath.com

11. The bar graph shows how much three students earned. Together they earned a total of $31.50. Student C earned $\frac{1}{2}$ as much as Student A. Find how much Student C earned.

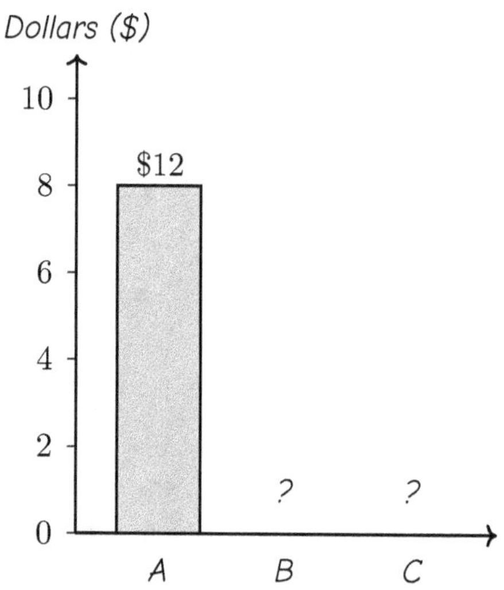

Student A earned $12, and Student C earned $\frac{1}{2}$ of Student A's earnings.

Your Answer:

12. Solve $7 - 2x < 1$.

(A) $x < 3$

(B) $x > 3$

(C) $x < -3$

(D) $x > -3$

13. Solve $-3x + 9 > 0$. Then graph the solution on the number line below by describing the circle type and shading direction.

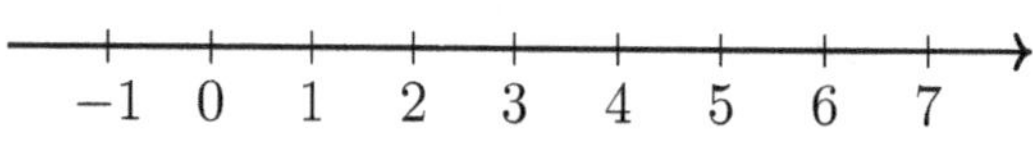

Your Answer:

Find more at
ViewMath.com/IN-Grade7

ViewMath.com

14. A rectangular swimming pool is 6 cm by 2.5 cm on a drawing with scale 1 cm = 8 m. What is the actual perimeter of the pool?

(A) 17 m

(B) 68 m

(C) 136 m

(D) 960 m

15. Which set of conditions produces more than one possible triangle?

(A) Sides 3, 4, 5 (SSS)

(B) Sides 6 and 8 with included angle 60° (SAS)

(C) Angles 40°, 60°, 80° (AAA)

(D) Angles 50°, 60° with included side 7 cm (ASA)

16. Two angles of a triangle are 72° and 53°. What is the third angle?

Your Answer

17. A hexagonal prism is sliced parallel to its base. What is the cross-section?

(A) Rectangle

(B) Triangle

(C) Hexagon

(D) Circle

18. Two lines intersect as shown. Find the values of a, b, and c.

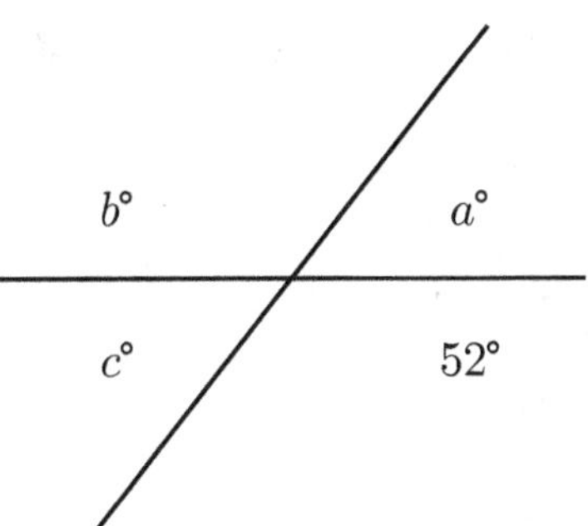

Your Answer:

19. Point $H(-3, 7)$ is reflected over the y-axis. What are the coordinates of H'?

(A) $(-3, -7)$

(B) $(3, 7)$

(C) $(3, -7)$

(D) $(7, -3)$

20. A pizza has a radius of 7 inches. What is the area of the pizza? Use $\pi \approx \frac{22}{7}$.

(A) $44\ in^2$

(B) $154\ in^2$

(C) $308\ in^2$

(D) $22\ in^2$

21. A composite shape is made of a rectangle that is 10 cm by 4 cm and a triangle with base 10 cm and height 3 cm attached to one side. What is the total area?

(A) $40\ cm^2$

(B) $55\ cm^2$

(C) $70\ cm^2$

(D) $120\ cm^2$

22. A cereal box is 30 cm tall, 20 cm wide, and 6 cm deep. What is its surface area?

(A) $1,440\ cm^2$

(B) $1,560\ cm^2$

(C) $1,800\ cm^2$

(D) $3,600\ cm^2$

23. A rectangular prism has a volume of 360 cm^3. Its length is 12 cm and its height is 5 cm. What is its width?

Your Answer

Find more at
ViewMath.com/IN-Grade7

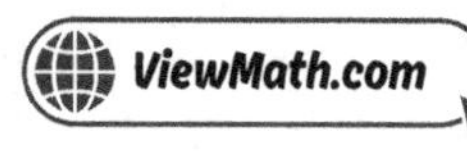

24. The table below shows the results of two samples taken from a population of 1,000 students.

	Sample 1 (50 students)	Sample 2 (50 students)
Prefer Math	18	22
Prefer Science	14	12
Prefer English	10	8
Prefer Art	8	8

Based on both samples, the best prediction for how many of the 1,000 students prefer math is:

(A) 180

(B) 220

(C) 400

(D) 360

25. Two histograms show that Group A's data is skewed left and Group B's data is symmetric. Which measure of center is most appropriate to compare them?

(A) Mean for both

(B) Median for both

(C) Mode for both

(D) Range for both

26. Data: 100, 200, 300, 400, 500. Find the median and the IQR.

Your Answer:

27. Which data display is best for showing how data changes over time?

(A) Circle graph

(B) Line graph

(C) Stem-and-leaf plot

(D) Box plot

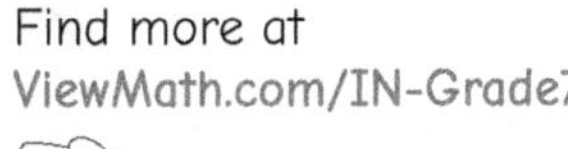

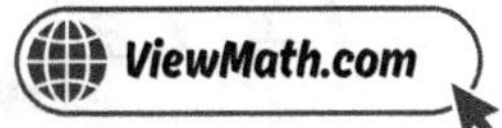

28. Look at the spinner below. Which probability model correctly represents this spinner?

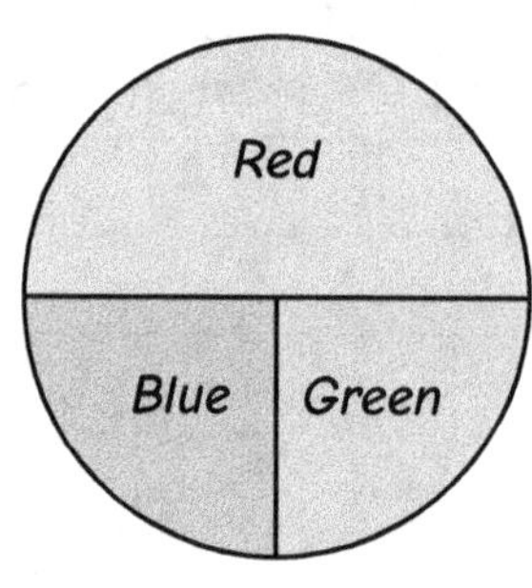

(A) $P(Red) = \frac{1}{3}$, $P(Blue) = \frac{1}{3}$, $P(Green) = \frac{1}{3}$

(B) $P(Red) = \frac{1}{2}$, $P(Blue) = \frac{1}{4}$, $P(Green) = \frac{1}{4}$

(C) $P(Red) = \frac{1}{4}$, $P(Blue) = \frac{1}{4}$, $P(Green) = \frac{1}{2}$

(D) $P(Red) = \frac{1}{2}$, $P(Blue) = \frac{1}{2}$, $P(Green) = 0$

29. Two dice are rolled. What is the probability that the product (not the sum) of the two numbers is 12?

(A) $\frac{2}{36}$

(B) $\frac{4}{36}$

(C) $\frac{3}{36}$

(D) $\frac{6}{36}$

30. A student simulates rolling two dice 36 times to see how often the sum is 7. They get a sum of 7 a total of 8 times. How does this compare to the expected number?

(A) Expected: 6; the simulation result (8) is slightly higher

(B) Expected: 6; the simulation result (8) is much higher

(C) Expected: 12; the simulation result (8) is lower

(D) Expected: 7; the simulation result (8) is exact

 # End of Practice Test 10 

Great job finishing the test!

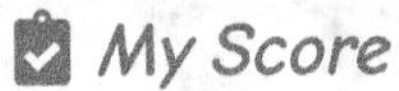 My Score

I got __________ out of 30 questions right.

*Check your answers in the **Answer Key** at the back of the book.*

💡 *Review any questions you missed. That's how we learn!*

📊 Check Your Score Online!

Visit **ViewMath Academy** to enter your answers and see which topics you need to review. You can also explore lessons, take quizzes, track your scores, and save your progress!

viewmath.com/score/7.1.IN.25

Or go to viewmath.com/score and enter code: 7.1.IN.25

Answer Key & Explanations

Answer Key

First try each test on your own, then check your work here.

✅ Practice Test 1 — Answer Key

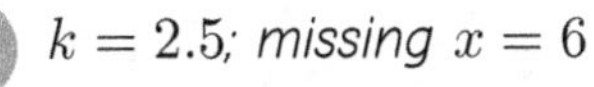 $k = 2.5$; missing $x = 6$  B    A    B    B    $\approx 4.2\%$  B    C

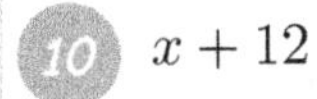 $-w + 4$ or $4 - w$  $x + 12$ 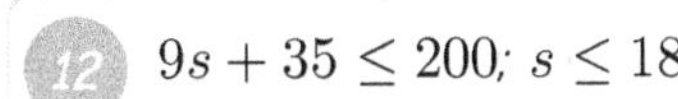 A   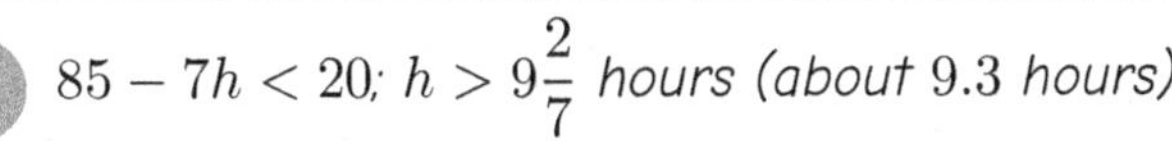 $9s + 35 \leq 200$; $s \leq 18$

 $85 - 7h < 20$; $h > 9\frac{2}{7}$ hours (about 9.3 hours)    C   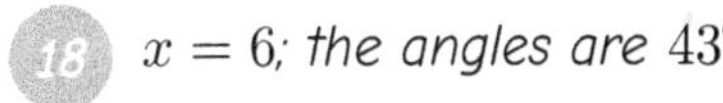 Exactly one    B

 Any two of: square, rectangle, triangle, pentagon, hexagon    $x = 6$; the angles are 43°    B

 B    B    A    C    B    C    B    B    0.4   29 B

30 48

💡 Time to Learn! 💡

Review the explanations below, **especially for the questions you missed**.

Understanding why each answer is correct builds stronger problem-solving skills.

Tip: Circle any questions you got wrong, then read their explanation carefully.

📖 Practice Test 1 — Detailed Explanations

1. $k = \frac{7.5}{3} = 2.5$. For the missing row: $15 = 2.5x$ so $x = 15 \div 2.5 = 6$.

2. Line P: $k = \frac{7}{5} = 1.4$. Line Q: $k = \frac{6}{6} = 1$. Line P is steeper and has the greater unit rate.

3. Worker A: $60 \div 5 = 12$ boxes/hr. Worker B: $45 \div 5 = 9$ boxes/hr. Worker A is faster by $12 - 9 = 3$ boxes/hr.

4. $\frac{n}{500} = \frac{68}{100}$. Cross-multiply: $100n = 34{,}000$. Divide: $n = 340$.

5. Commission: $6{,}000 \times 0.05 = \$300$. Total: $400 + 300 = \$700$.

6. $\frac{|5 - 4.8|}{4.8} \times 100 = \frac{0.2}{4.8} \times 100 \approx 4.17 \approx 4.2\%$.

7. Point P is at -4 and Q is at 4. They are the same distance from 0 on opposite sides, so they are opposites.

8. Adding 0 to any number gives that number. $0 + (-7) = -7$.

9. Combine variable terms: $-5w + 3w + w = -w$. Combine constants: $12 - 8 = 4$. Result: $-w + 4$.

10. Expand: $6x + 12$. Subtract $5x$: $6x + 12 - 5x = x + 12$.

11. Equation: $-3.5 + 4t = 8.5$. Add 3.5: $4t = 12$. Divide by 4: $t = 3$.

12. Total cost: $9s + 35 \leq 200$. Subtract 35: $9s \leq 165$. Divide by 9: $s \leq 18.\overline{3}$. Since s must be a whole number, $s \leq 18$.

13. $85 - 7h < 20$. Subtract 85: $-7h < -65$. Divide by -7 and flip: $h > \frac{65}{7} = 9\frac{2}{7}$.

14. Multiply the map distance by the scale factor. $6 \times 25 = 150$ km.

Find more at
ViewMath.com/IN-Grade7

15 ASA (two angles and the included side) fully determine a unique triangle. The third angle would be $180° - 45° - 45° = 90°$.

16 Two sides and the angle between them is the SAS (Side-Angle-Side) condition.

17 A cube can be sliced to produce squares, rectangles, triangles, pentagons, or hexagons depending on the angle and position of the cut.

18 Vertical angles are equal: $7x + 1 = 5x + 13$. Subtract $5x$: $2x + 1 = 13$. Subtract 1: $2x = 12$. Divide: $x = 6$. Angle: $7(6) + 1 = 43°$.

19 The x-coordinate stayed the same. The y-coordinate went from 7 to 3, a change of -4. That is down 4.

20 $r = 7$ cm. $A = \frac{22}{7} \times 49 = 154$ cm^2.

21 Rectangle: $12 \times 5 = 60$ cm^2. Triangle: $\frac{1}{2} \times 4 \times 5 = 10$ cm^2. Remaining: $60 - 10 = 50$ cm^2.

22 Two triangles: $2 \times \frac{1}{2} \times 6 \times 4 = 24$ cm^2. Three rectangles: $6 \times 10 + 5 \times 10 + 5 \times 10 = 60 + 50 + 50 = 160$ cm^2. Total: $24 + 160 = 184$ cm^2.

23 $V = 50 \times 30 \times 40 = 60{,}000$ cm^3.

24 $\frac{15}{60} = \frac{1}{4} = 25\%$. So 25% of $600 = 150$.

25 A box plot shows the quartiles (Q_1 and Q_3), so you can read the IQR directly as $Q_3 - Q_1$.

26 Company X has smaller IQR (1.5 vs. 4) and range (4 vs. 8), meaning delivery times are more predictable.

27 Boys: $\frac{80+85+90}{3} = 85$. Girls: $\frac{75+90+95}{3} = \frac{260}{3} \approx 86.7$. Girls are slightly higher.

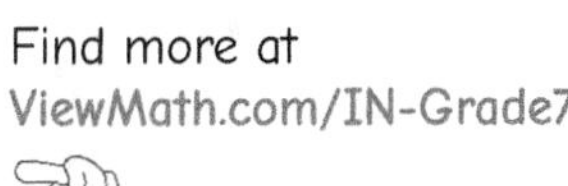

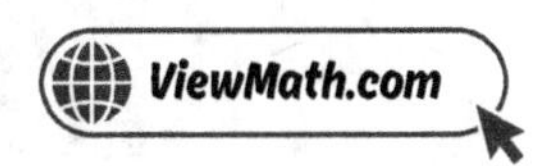

28 $P(Z) = 1 - 0.15 - 0.45 = 0.40.$

29 $P(spinner = 3) = \frac{1}{4}, P(die = 3) = \frac{1}{6}. \ P = \frac{1}{4} \times \frac{1}{6} = \frac{1}{24}.$

30 $80 \times 0.60 = 48.$

✅ Practice Test 2 — Answer Key

1 B

2 0.75 birdhouses per hour; 7.5 birdhouses in 10 hours

3 $37\frac{1}{2}$ cups

4 A

5 D

6 C

7 D

8 -5

9 A

10 $6a - 9$

11 B

12 $x \leq -5$

13 $s \geq 87$; closed circle at 87, shade right

14 The area does not change

15 C

16 B

17 A

18 B

19 $P'(-6, -2)$

20 B

21 B

22 156 cm^2

23 C

24 4%

25 B

26 City A: mean = 66.6°F, range = 10°F. City B: mean = 68.2°F, range = 18°F. City A is more consistent.

27 B

28 B

29 C

30 B

💡 Time to Learn! 💡

Review the explanations below, **especially for the questions you missed.**

Understanding why each answer is correct builds stronger problem-solving skills.

Tip: Circle any questions you got wrong, then read their explanation carefully.

📖 Practice Test 2 — Detailed Explanations

Find more at
ViewMath.com/IN-Grade7

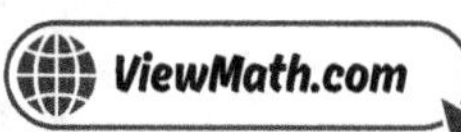

1. $k = \frac{54}{2} = 27$, $\frac{135}{5} = 27$, $\frac{216}{8} = 27$. The car gets 27 miles per gallon.

2. $k = \frac{3}{4} = 0.75$. In 10 hours: $0.75 \times 10 = 7.5$ birdhouses.

3. $k = \frac{7.5}{3} = 2.5$ cups per batch. For 15 batches: $2.5 \times 15 = 37.5 = 37\frac{1}{2}$ cups.

4. Part (42) over whole (w) equals percent (60) over 100: $\frac{42}{w} = \frac{60}{100}$.

5. $17.20 \div 86 = 0.20 = 20\%$.

6. A: $\frac{|180-200|}{200} = 10\%$. B: $\frac{|220-200|}{200} = 10\%$. Both have 10% error.

7. Absolute value is the distance from 0. Since -9 is 9 units from 0, $|-9| = 9$.

8. Group positives: $5 + 3 = 8$. Group negatives: $(-11) + (-2) = -13$. Combine: $8 + (-13) = -5$.

9. Combine variable terms: $-4x + 7x = 3x$. Combine constants: $9 - 3 = 6$. Result: $3x + 6$.

10. $\frac{3}{4} \times 8a = 6a$ and $\frac{3}{4} \times (-12) = -9$. Result: $6a - 9$.

11. Subtract 3: $\frac{5x}{2} = 15$. Multiply by 2: $5x = 30$. Divide by 5: $x = 6$. Check: $\frac{5(6)}{2} + 3 = 15 + 3 = 18$ ✓.

12. Subtract 8: $-3x \geq 15$. Divide by -3 and flip: $x \leq -5$.

13. $\frac{78+85+90+s}{4} \geq 85$. Multiply by 4: $253 + s \geq 340$. Subtract 253: $s \geq 87$. Closed circle at 87, shade right.

14. The actual room stays the same. Changing the drawing scale does not change the real room's area. The area of the real room is always $12 \times 12 = 144 \ m^2$.

Find more at
ViewMath.com/IN-Grade7

15 $6 + 8 = 14$, which is not greater than 14. The sum must be strictly greater, so no triangle can be formed.

16 $180° - 35° - 75° = 70°$.

17 A vertical cut through a cube perpendicular to the base typically produces a rectangle. If the cut goes through the center parallel to a face, it can be a square.

18 Let the smaller be x. Then $x + (x + 25) = 180$. Combine: $2x + 25 = 180$. Solve: $2x = 155$, $x = 77.5°$. The other: $77.5 + 25 = 102.5°$.

19 Over the x-axis: keep x, negate y. $(-6, 2) \to (-6, -2)$.

20 $r^2 = A \div \pi = 78.5 \div 3.14 = 25$. So $r = 5$ cm.

21 Rectangle: $25 \times 10 = 250$ m^2. Semicircle: $\frac{1}{2} \times 3.14 \times 5^2 = \frac{1}{2} \times 78.5 = 39.25$ m^2. Total: $250 + 39.25 = 289.25$ m^2.

22 Two triangles: $2 \times \frac{1}{2} \times 3 \times 4 = 12$ cm^2. Three rectangles: $3 \times 12 + 4 \times 12 + 5 \times 12 = 36 + 48 + 60 = 144$ cm^2. Total: $12 + 144 = 156$ cm^2.

23 $B = \frac{1}{2} \times 8 \times 3 = 12$ cm^2. $V = Bh = 12 \times 10 = 120$ cm^3.

24 $\frac{8}{200} = 0.04 = 4\%$.

25 Class 1 center ≈ 10. Class 2 center ≈ 13. Difference ≈ 3 push-ups. Class 2 is higher.

26 City A: $(62 + 64 + 64 + 66 + 66 + 66 + 68 + 68 + 70 + 72) \div 10 = 666 \div 10 = 66.6$. Range $= 72 - 62 = 10$. City B: $(58 + 60 + 64 + 66 + 68 + 70 + 72 + 74 + 74 + 76) \div 10 = 682 \div 10 = 68.2$. Range $= 76 - 58 = 18$. City A has the smaller range.

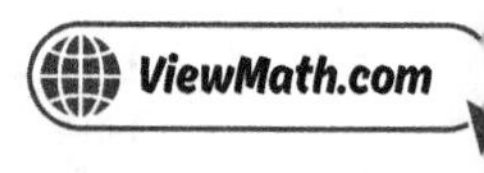

27 A frequency table shows each category (or interval) and the count of data values in it.

28 Each color should appear about 50 times (200×0.25). The results $(55, 48, 52, 45)$ are all close to 50, supporting the model.

29 Outcomes with at least one tail: HT, TH, TT — that is 3 out of 4. $P = \frac{3}{4}$.

30 With 4 equal sections, each should appear about 25 times in 100 spins. The results $(26, 24, 28, 22)$ are all close to 25, suggesting the spinner is fair.

☑ Practice Test 3 — Answer Key

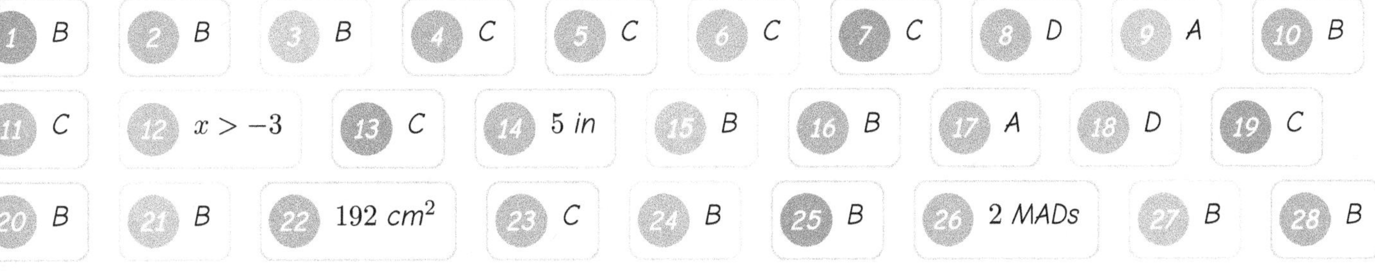

1 B	2 B	3 B	4 C	5 C
6 C	7 C	8 D	9 A	10 B
11 C	12 $x > -3$	13 C	14 5 in	15 B
16 B	17 A	18 D	19 C	20 B
21 B	22 192 cm²	23 C	24 B	25 B
26 2 MADs	27 B	28 B	29 C	30 A

💡 Time to Learn! 💡

Review the explanations below, **especially for the questions you missed**.

Understanding why each answer is correct builds stronger problem-solving skills.

Tip: Circle any questions you got wrong, then read their explanation carefully.

📖 Practice Test 3 — Detailed Explanations

Find more at
ViewMath.com/IN-Grade7

ViewMath.com

1. If $k = \frac{3}{4}$, then $y = \frac{3}{4}x$. For $x = 8$: $y = \frac{3}{4} \times 8 = 6$. So $(8, 6)$ works.

2. A steeper line means a higher slope, which equals a greater unit rate (constant of proportionality).

3. Store A: $7.50 \div 3 = \$2.50$ each. Store B: $11.25 \div 5 = \$2.25$ each. Store B is cheaper.

4. Proportion: $\frac{n}{75} = \frac{40}{100}$ gives $n = 30$. Equation: $n = 0.40 \times 75 = 30$. Both give 30.

5. 10% of $\$78 = \$7.80$. Half of that (5%) $= \$3.90$. Add: $7.80 + 3.90 = \$11.70$.

6. $\frac{|4.5 - 5|}{5} \times 100 = \frac{0.5}{5} \times 100 = 10\%$.

7. Both $|10| = 10$ and $|-10| = 10$. Two numbers have the same absolute value: the number and its opposite.

8. A number plus its opposite (additive inverse) always equals 0.

9. Combine: $\frac{3}{4}n - \frac{1}{4}n = \frac{2}{4}n = \frac{1}{2}n$. Add the constant: $\frac{1}{2}n + 5$.

10. $\frac{2}{3} \times 9x = 6x$ and $\frac{2}{3} \times 6 = 4$. Result: $6x + 4$.

11. Subtract 4.5: $-1.5x = -7.5$. Divide by -1.5: $x = 5$. Check: $-1.5(5) + 4.5 = -7.5 + 4.5 = -3$ ✓.

12. Add 4: $-7x < 21$. Divide by -7 and flip: $x > -3$.

13. Divide by -2 and flip: $x \geq -5$. Closed circle at -5, shade right.

14. $45 \div 9 = 5$ in.

Find more at
ViewMath.com/IN-Grade7

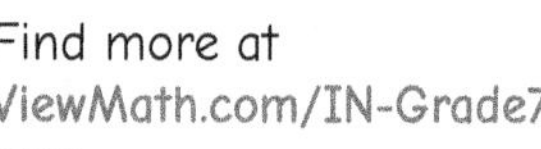

15 AAA with a specified side length fixes the size. A 60°-60°-60° triangle is equilateral, so all sides are 4 cm. Exactly one triangle.

16 Third angle: $180° - 25° - 25° = 130°$. Two equal angles means two equal sides, making it isosceles.

17 Cut A (horizontal, parallel to the base) produces a circle. Cut B (vertical through the apex) produces a triangle.

18 Let the supplement be x. The angle is $4x$. Sum: $x + 4x = 180$. So $5x = 180$ and $x = 36°$. The angle: $4(36) = 144°$.

19 Over x-axis: $(3, -5) \to (3, 5)$. Over y-axis: $(3, 5) \to (-3, 5)$.

20 $r = 8 \div 2 = 4$ ft. $A = \pi r^2 = 3.14 \times 16 = 50.24$ ft^2.

21 Triangle: $\frac{1}{2} \times 6 \times 8 = 24$ cm^2. Rectangle: $8 \times 3 = 24$ cm^2. Total: $24 + 24 = 48$ cm^2.

22 The prism is $10 \times 4 \times 4$. $SA = 2(10 \times 4) + 2(10 \times 4) + 2(4 \times 4) = 80 + 80 + 32 = 192$ cm^2.

23 $B = \frac{1}{2}(4 + 6) \times 3 = 15$ cm^2. $V = 15 \times 10 = 150$ cm^3.

24 Larger random samples better represent the population and give more reliable results.

25 Difference in medians: $88 - 82 = 6$. In MADs: $\frac{6}{4} = 1.5$ MADs.

26 Difference: $57 - 45 = 12$. In MADs: $\frac{12}{6} = 2$.

27 Frequency tables can list categories (favorite color) or numerical intervals (test scores 70–79).

Find more at
ViewMath.com/IN-Grade7

28. The probability of drawing red $(\frac{4}{5})$ is different from drawing blue $(\frac{1}{5})$, so this is a non-uniform model.

29. The only way to get a sum of 2 is $(1, 1)$. $P = \frac{1}{36}$.

30. $P = \frac{6}{40} = 0.15$. The theoretical probability is $\frac{1}{8} = 0.125$, close to the simulation result.

✔ Practice Test 4 — Answer Key

1 B	2 $112.50	3 B	4 36	5 C	6 C	7 C	8 15 feet	9 C
10 C	11 B	12 $\frac{x}{4} - 3 \geq 2$; $x \geq 20$	13 A	14 $1\ cm = 30\ m$	15 D	16 C	17 C
18 A	19 A	20 D	21 B	22 C	23 B

24. Random variation (or the sample may be biased)   25 C   26 B   27 C   28 C   29 $\frac{3}{8}$

30 B

💡 Time to Learn! 💡

Review the explanations below, **especially for the questions you missed**.

Understanding why each answer is correct builds stronger problem-solving skills.

Tip: Circle any questions you got wrong, then read their explanation carefully.

📖 Practice Test 4 — Detailed Explanations

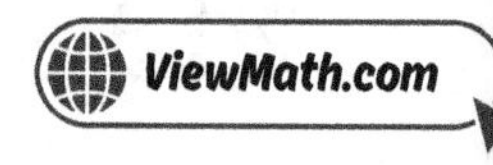

1. $k = \frac{y}{x} = \frac{12}{3} = 4$. Check: $\frac{20}{5} = 4$ and $\frac{32}{8} = 4$.

2. $k = \frac{67.50}{3} = 22.50$ per hour. Cost for 5 hours: $22.50 \times 5 = \$112.50$.

3. Factory 1: $168 \div 7 = 24$ items/hr. Factory 2: $200 \div 8 = 25$ items/hr. Factory 2 is more productive.

4. $\frac{n}{80} = \frac{45}{100}$. Cross-multiply: $100n = 3600$. Divide: $n = 36$ tomato plants.

5. $15\% = 0.15$. Tip: $40 \times 0.15 = \$6.00$.

6. Absolute value ensures the percent error is positive whether the estimate was too high or too low.

7. Absolute value measures distance from 0, which is always ≥ 0. It can never be negative.

8. $(-30) + 45 = 15$. She is now 15 feet above sea level.

9. Combine variable terms: $8b + 2b - b = 9b$. Combine constants: $-3 + 5 = 2$. Simplified: $9b + 2$, which has 2 terms.

10. $5 \times 2a = 10a$ and $5 \times (-3) = -15$. Result: $10a - 15$.

11. Multiply every term by 6: $3n + 2n = 60$. Combine: $5n = 60$. Divide by 5: $n = 12$. Check: $\frac{12}{2} + \frac{12}{3} = 6 + 4 = 10$ ✓.

12. $\frac{x}{4} - 3 \geq 2$. Add 3: $\frac{x}{4} \geq 5$. Multiply by 4: $x \geq 20$.

13. Solve: $2x > 8$, so $x > 4$. This is an open circle at 4, shade right — Graph A.

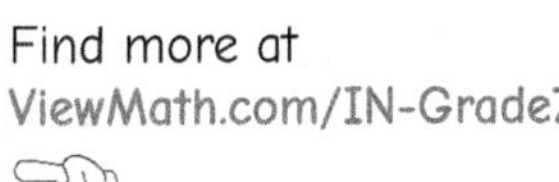
Find more at
ViewMath.com/IN-Grade7

14 $360 \div 12 = 30$. The scale is 1 cm = 30 m.

15 Three angles (AAA) determine the shape but not the size. You can draw infinitely many similar triangles of different sizes with these angles.

16 Both triangles have the same angles but different side lengths. AAA determines shape but not size, so many triangles are possible.

17 A rectangular prism has flat faces and straight edges. Its cross-sections can be triangles, rectangles, pentagons, or hexagons, but never circles (since it has no curved surfaces).

18 Vertical angles are equal (130° and 130°). Adjacent angles are supplementary: $180° - 130° = 50°$. The four angles are 130°, 50°, 130°, 50°.

19 $C(5, 3)$: left 3 means $5 - 3 = 2$, up 4 means $3 + 4 = 7$. So $C'(2, 7)$.

20 $A = \pi r^2 = 3.14 \times 100 = 314 \ cm^2$.

21 Square: $8^2 = 64 \ cm^2$. Semicircle: $\frac{1}{2} \times \pi r^2 = \frac{1}{2} \times 3.14 \times 16 = 25.12 \ cm^2$. Total: $64 + 25.12 = 89.12 \ cm^2$.

22 The three pairs of faces have areas $7 \times 4 = 28$, $7 \times 3 = 21$, and $4 \times 3 = 12 \ m^2$. The largest is $28 \ m^2$.

23 Bottom row: 5 cubes. Second row: 3 cubes. Third row: 3 cubes. Total: $5 + 3 + 3 = 11 \ cm^3$.

24 Even a well-chosen random sample may differ slightly from the population due to chance.

25 A complete comparison requires looking at both center (where the data clusters) and spread (how spread out it is).

26 Difference $= \$620 - \$500 = \$120$. In MADs: $\frac{120}{40} = 3$ MADs.

Find more at
ViewMath.com/IN-Grade7

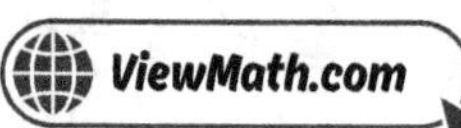

27 *Week 1–2: 3 cm. Week 2–3: 4 cm. Week 3–4: 5 cm. The most growth was Week 3 to 4.*

28 $P(tails) = 1 - 0.60 = 0.40$. *Since heads and tails have different probabilities, this is a non-uniform model.*

29 *Outcomes with exactly 2 tails: HTT, THT, TTH — that is 3 out of 8.* $P = \frac{3}{8}$.

30 *Each coin flip has a 50% chance of heads (pass). Flipping 3 coins and counting exactly 2 heads simulates exactly 2 passes.*

☑ Practice Test 5 — Answer Key

1 $k = 70$ widgets per hour   2 B   3 B   4 B   5 C   6 $\approx 7.1\%$ 7 C

8 C   9 C   10 $3n - 9$ or $3(n - 3)$ 11 $x = 12$ 12 B

13 $x > 2$; open circle at 2, shade right   14 B   15 D   16 AAA   17 B   18 B   19 B

20 C   21 $52\ \text{cm}^2$ 22 C   23 B   24 B   25 A   26 B

27 Total $= 14$ in.; March had the most (5 in.)   28 C   29 $\frac{1}{9}$ 30 C

💡 Time to Learn! 💡

*Review the explanations below, **especially for the questions you missed.***

Understanding why each answer is correct builds stronger problem-solving skills.

***Tip:** Circle any questions you got wrong, then read their explanation carefully.*

Find more at
ViewMath.com/IN-Grade7

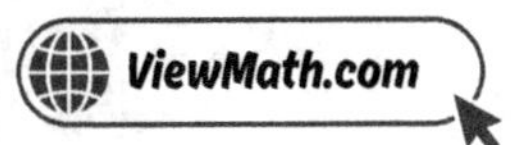

📖 Practice Test 5 — Detailed Explanations

1. $k = \frac{175}{2.5} = 70$ widgets per hour.

2. Unit rate $= k = \frac{18}{3} = 6$.

3. 12-oz: $3.60 \div 12 = \$0.30/oz$. 20-oz: $5.40 \div 20 = \$0.27/oz$. The 20-oz box is the better value.

4. $\frac{n}{80} = \frac{37.5}{100}$. Cross-multiply: $100n = 3000$. Divide: $n = 30$.

5. $250{,}000 \times 0.03 = \$7{,}500$.

6. $\frac{|45-42|}{42} \times 100 = \frac{3}{42} \times 100 \approx 7.14 \approx 7.1\%$.

7. On a number line, -7 is to the left of -4, so $-7 < -4$.

8. $(-9) + n = -2$ means $n = -2 - (-9) = -2 + 9 = 7$.

9. Combine variable terms: $-3x + 5x = 2x$. Combine constants: $10 - 4 = 6$. Result: $2x + 6$.

10. Expand: $5n - 15 - 2n + 6$. Combine: $(5n - 2n) + (-15 + 6) = 3n - 9$.

11. Add 2: $\frac{3x}{4} = 9$. Multiply by 4: $3x = 36$. Divide by 3: $x = 12$. Check: $\frac{3(12)}{4} - 2 = 9 - 2 = 7$ ✓.

12. Add 10: $5x \geq 35$. Divide by 5: $x \geq 7$.

13. Subtract 10: $-4x < -8$. Divide by -4 and flip: $x > 2$. Open circle at 2, shade right.

Find more at
ViewMath.com/IN-Grade7

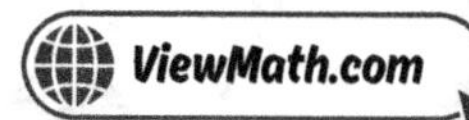

14 $3.2 \times 50 = 160$ km.

15 The sides must add to 12. Many dimension pairs work: 1×11, 2×10, 3×9, etc. Perimeter alone does not fix the rectangle.

16 Three angles determine the shape but not the size. Infinitely many similar triangles can share the same three angle measures.

17 The shape of a cross-section depends on where you cut and at what angle. Different cuts through the same figure can produce different shapes.

18 Vertical angles are always equal. If one is 72°, the vertical angle is also 72°.

19 Translations, reflections, and rotations are all rigid motions — they preserve size and shape. Reflections reverse orientation, and rotations change direction, but all keep the figure congruent.

20 $A = \pi r^2 = 3.14 \times 25 = 78.5$ m^2.

21 Top: $10 \times 2 = 20$ cm^2. Bottom: $10 \times 2 = 20$ cm^2. Middle: $2 \times 6 = 12$ cm^2. Total: $20 + 20 + 12 = 52$ cm^2.

22 A rectangular prism has 6 faces: top, bottom, front, back, left, and right.

23 Space volume: $6 \times 4 \times 3 = 72$ ft^3. Box volume: $2 \times 2 \times 3 = 12$ ft^3. Number: $72 \div 12 = 6$ boxes.

24 The population is the entire group you want to study — all 800 students.

25 A smaller MAD means scores are closer to the mean, so Team A is more consistent.

26 Same mean means same center. Different MADs mean one group's data is more spread out than the other's.

27 $2 + 3 + 5 + 4 = 14$ inches. March has the highest value at 5 in.

28 There are 2 B tiles out of 5 total. $P(B) = \frac{2}{5}$.

29 Products of 6: $(1,6), (2,3), (3,2), (6,1) - 4$ outcomes. $P = \frac{4}{36} = \frac{1}{9}$.

30 Before choosing a model or running trials, you must first identify what event you're simulating and what the possible outcomes are.

✅ Practice Test 6 — Answer Key

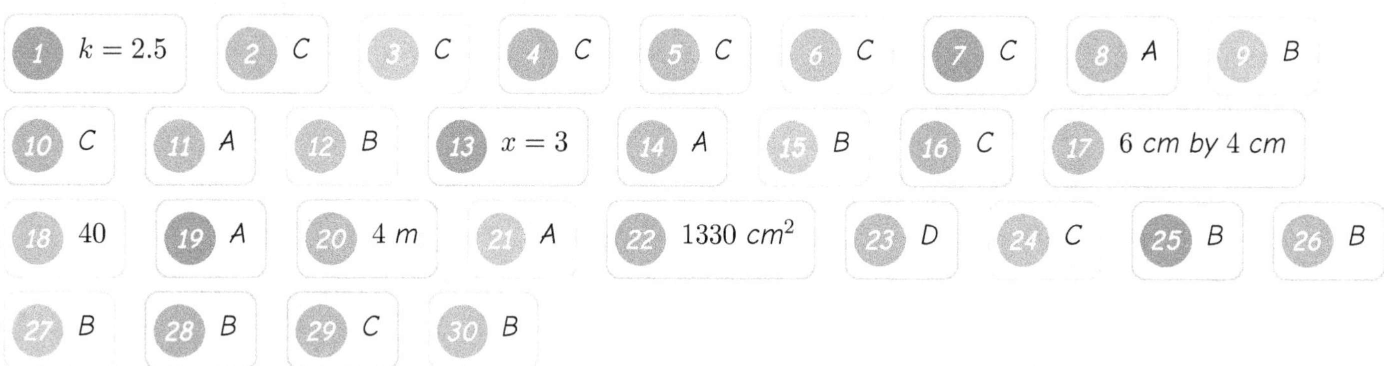

| 1 | $k = 2.5$ | 2 | C | 3 | C | 4 | C | 5 | C | 6 | C | 7 | C | 8 | A | 9 | B |

| 10 | C | 11 | A | 12 | B | 13 | $x = 3$ | 14 | A | 15 | B | 16 | C | 17 | 6 cm by 4 cm |

| 18 | 40 | 19 | A | 20 | 4 m | 21 | A | 22 | 1330 cm^2 | 23 | D | 24 | C | 25 | B | 26 | B |

| 27 | B | 28 | B | 29 | C | 30 | B |

💡 Time to Learn! 💡

Review the explanations below, **especially for the questions you missed**.

Understanding why each answer is correct builds stronger problem-solving skills.

Tip: Circle any questions you got wrong, then read their explanation carefully.

📖 Practice Test 6 — Detailed Explanations

Find more at
ViewMath.com/IN-Grade7

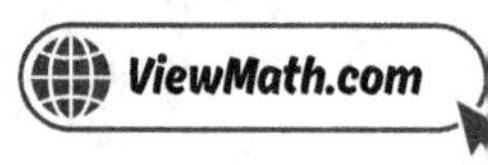
ViewMath.com

1. $k = \frac{y}{x} = \frac{15}{6} = 2.5$.

2. $k = \frac{10}{4} = 2.5$. When $x = 7$: $y = 2.5 \times 7 = 17.5$.

3. $\frac{3}{87} = \frac{x}{435}$. Cross-multiply: $87x = 1{,}305$, so $x = 15$ gallons.

4. $\frac{24}{80} = \frac{p}{100}$. Cross-multiply: $80p = 2400$. Divide: $p = 30\%$.

5. Planning fee: $4{,}000 \times 0.10 = \$400$. Coordination fee: $4{,}000 \times 0.03 = \$120$. Total: $400 + 120 = \$520$.

6. $\frac{|30-28|}{28} \times 100 = \frac{2}{28} \times 100 \approx 7.1\%$.

7. The opposite of 4 is -4. The opposite of -4 is 4. Taking the opposite twice returns you to the original number.

8. $4 + (-10) = -6$. Check the others: $(-3) + (-9) = -12$; $8 + (-2) = 6$; $(-1) + (-4) = -5$.

9. Combine $9a - 4a = 5a$. The constant 6 stays. Result: $5a + 6$.

10. $\frac{1}{2} \times 8x = 4x$ and $\frac{1}{2} \times (-6) = -3$. Result: $4x - 3$.

11. Subtract 5.6: $-0.2n = -1.6$. Divide by -0.2: $n = 8$. Check: $-0.2(8) + 5.6 = -1.6 + 5.6 = 4$ ✓.

12. An open circle at 5 with shading to the right means $x > 5$ (not including 5).

13. Add 3: $7x \geq 21$. Divide by 7: $x \geq 3$. The smallest integer solution is 3.

14. Divide the actual length by the scale factor: $180 \div 24 = 7.5$ inches.

Find more at
ViewMath.com/IN-Grade7

15 A square with area 36 cm^2 has side 6 cm. All squares with side 6 cm are congruent, so exactly one square satisfies this.

16 Although $90 + 100 + (-10) = 180$, every angle in a triangle must be greater than $0°$. A negative angle is impossible.

17 A cut parallel to the base of a rectangular prism produces a rectangle identical to the base: 6 cm by 4 cm.

18 Angles on a line sum to $180°$: $50 + 2x + 10 + x = 180$. Combine: $3x + 60 = 180$. Solve: $3x = 120$, $x = 40$.

19 Left 3: $-4 - 3 = -7$. Up 5: $-1 + 5 = 4$. So $E'(-7, 4)$.

20 $r^2 = 12.56 \div 3.14 = 4$. So $r = 2$ m and $d = 4$ m.

21 Room: $14 \times 10 = 140$ ft^2. Closet: $3 \times 4 = 12$ ft^2. Remaining: $140 - 12 = 128$ ft^2.

22 $SA = 2(25 \times 18) + 2(25 \times 5) + 2(18 \times 5) = 900 + 250 + 180 = 1330$ cm^2.

23 First: $8 \times 5 \times 4 = 160$ in^3. Second: $10 \times 4 \times 4 = 160$ in^3. They have the same volume.

24 In a random sample, every member of the population has an equal chance of being chosen.

25 Dot plots show every data point, which is useful for small data sets where individual values matter.

26 There are 7 values. The middle value (4th) is 20.

27 A double bar graph is designed to compare two groups for the same categories by placing their bars side by side.

Find more at
ViewMath.com/IN-Grade7

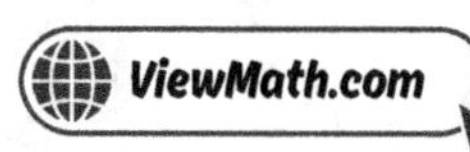

28 The probabilities are $0.4, 0.3, 0.2, 0.1$ — all different, so this is non-uniform. They sum to 1.0, making it valid.

29 $P(\text{tails}) = \frac{1}{2}$ and $P(\text{even}) = \frac{3}{6} = \frac{1}{2}$. $P = \frac{1}{2} \times \frac{1}{2} = \frac{1}{4}$. Or: 3 favorable out of 12 total outcomes.

30 A simulation uses random numbers, coins, dice, or spinners to model a real-world event and estimate probabilities.

✅ Practice Test 7 — Answer Key

1 $y = 36$	2 D	3 7 hours	4 D	5 $81.60	6 C	7 -3	8 B
9 B	10 D	11 $p = \$60$	12 A	13 A	14 C	15 No	16 D	17 C
18 C	19 A	20 39.25 cm^2	21 C	22 C	23 180 cm^3	24 B	25 Class A
26 B	27 Mean = 45 visitors; Range = 30; Friday was busiest (60 visitors)	28 B	29 B	30 C

💡 Time to Learn! 💡

Review the explanations below, **especially for the questions you missed**.

Understanding why each answer is correct builds stronger problem-solving skills.

Tip: Circle any questions you got wrong, then read their explanation carefully.

📖 Practice Test 7 — Detailed Explanations

1 $k = \frac{20}{5} = 4$. So $y = 4 \times 9 = 36$.

Find more at
ViewMath.com/IN-Grade7

2 A proportional relationship always graphs as a straight line through the origin. It cannot be curved.

3 Rate: $42 \div 3 = 14$ mph. Time: $98 \div 14 = 7$ hours.

4 $\frac{135}{w} = \frac{54}{100}$. Cross-multiply: $54w = 13{,}500$. Divide: $w = 250$.

5 Fee: $85 \times 0.04 = \$3.40$. Seller receives: $85 - 3.40 = \$81.60$.

6 $\frac{|50-40|}{40} \times 100 = \frac{10}{40} \times 100 = 25\%$.

7 Moving 5 units left from 2: $2 - 5 = -3$.

8 $(-20) + 8 = -12$. Rising from a negative position adds a positive: $20 - 8 = 12$, keep negative.

9 Like terms have the same variable raised to the same power. $5m$ and $-2m$ both have the variable m to the first power.

10 Expand: $7a - 21$. Subtract $2a$: $7a - 21 - 2a = 5a - 21$.

11 After $\frac{1}{4}$ markdown: $p - \frac{p}{4} = \frac{3p}{4}$. After subtracting \$8: $\frac{3p}{4} - 8 = 37$. Add 8: $\frac{3p}{4} = 45$. Multiply by $\frac{4}{3}$: $p = 60$.

12 Current height plus growth: $42 + g > 48$.

13 Subtract 3: $4x \leq 16$. Divide by 4: $x \leq 4$. Closed circle at 4, shade left.

14 A scale drawing keeps all lengths proportional using the same scale factor. Angles are preserved, and area scales by k^2, not k.

Find more at
ViewMath.com/IN-Grade7

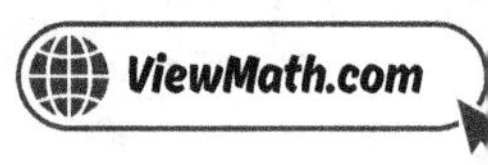

15 $6 + 3 = 9 < 10$. *The two shorter sides cannot span the longest side. The triangle inequality is not satisfied.*

16 *AAA determines the shape but not the size. Infinitely many similar triangles of different sizes share these angle measures.*

17 *A horizontal cut parallel to the base of a rectangular prism produces a rectangle the same shape as the base.*

18 *If two equal angles sum to* $180°$*, then each is* $180 \div 2 = 90°$*.*

19 *90° clockwise:* $(1, 4) \to (4, -1)$*. Reflect over* x*-axis:* $(4, -1) \to (4, 1)$*. Final:* $(4, 1)$*.*

20 $r = 5$ *cm. Full circle:* $3.14 \times 25 = 78.5$ *cm^2. Semicircle:* $78.5 \div 2 = 39.25$ *cm^2.*

21 *Horizontal:* $12 \times 2 = 24$ *cm^2. Vertical:* $2 \times 8 = 16$ *cm^2. Total:* $24 + 16 = 40$ *cm^2.*

22 *Surface area is the total area covering the outside of a 3D object — the sum of the areas of all its faces.*

23 $V = 9 \times 5 \times 4 = 180$ *cm^3.*

24 *Selecting every 10th student from the school roster is systematic and gives all students a chance, making it the least biased.*

25 *Class A has a smaller IQR (6 vs. 14), meaning the middle 50% of scores are more tightly clustered.*

26 *A wider range indicates more variability, not better performance. Class A has a higher median and is more consistent (smaller IQR and range).*

27 *Sum:* $40 + 35 + 50 + 45 + 60 + 55 + 30 = 315$*. Mean:* $\frac{315}{7} = 45$*. Range:* $60 - 30 = 30$*. Friday had the most visitors (60).*

Find more at
ViewMath.com/IN-Grade7

28 A fair number cube has 6 equally likely outcomes, making it a uniform model.

29 $P(\text{heads}) = \frac{1}{2}$. Numbers greater than 4: $5, 6$, so $P = \frac{2}{6} = \frac{1}{3}$. $P = \frac{1}{2} \times \frac{1}{3} = \frac{1}{6}$.

30 A spinner with 4 equal sections has $\frac{1}{4} = 25\%$ chance for each section. Assigning one section as "win" gives 25%.

✅ Practice Test 8 — Answer Key

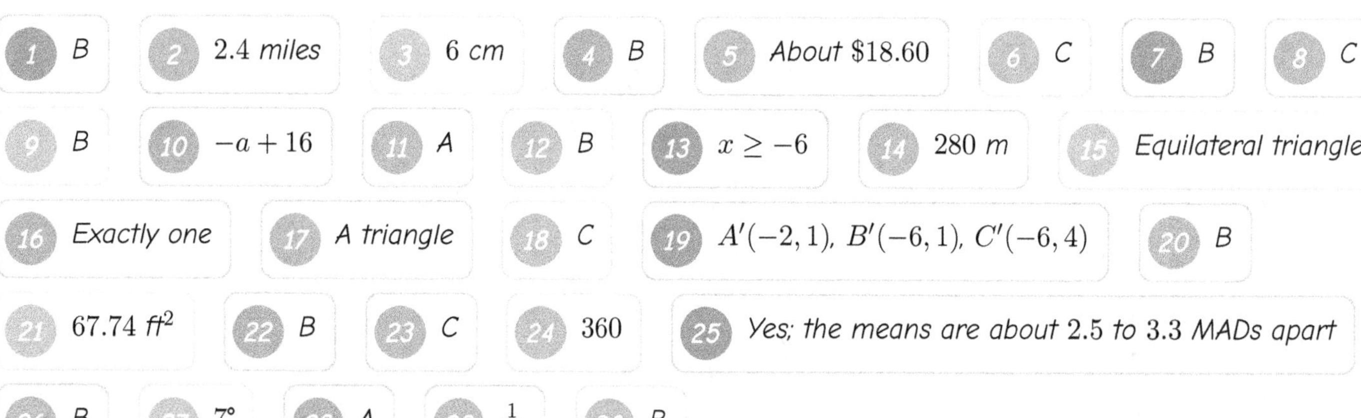

1 B	**2** 2.4 miles	**3** 6 cm	**4** B
5 About \$18.60	**6** C	**7** B	**8** C
9 B	**10** $-a + 16$	**11** A	**12** B
13 $x \geq -6$	**14** 280 m	**15** Equilateral triangle	**16** Exactly one
17 A triangle	**18** C	**19** $A'(-2, 1)$, $B'(-6, 1)$, $C'(-6, 4)$	**20** B
21 67.74 ft^2	**22** B	**23** C	**24** 360
25 Yes; the means are about 2.5 to 3.3 MADs apart	**26** B	**27** 7°	**28** A
29 $\frac{1}{12}$	**30** B		

💡 Time to Learn! 💡

Review the explanations below, **especially for the questions you missed**.

Understanding why each answer is correct builds stronger problem-solving skills.

Tip: Circle any questions you got wrong, then read their explanation carefully.

📖 Practice Test 8 — Detailed Explanations

Find more at
ViewMath.com/IN-Grade7

🌐 ViewMath.com

1. $k = \frac{150}{2.5} = 60$ *miles per hour.*

2. *Ava:* $k = \frac{6}{5} = 1.2$ *mph; at 6 hr:* $1.2 \times 6 = 7.2$ *mi. Ben:* $k = \frac{4}{5} = 0.8$ *mph; at 6 hr:* $0.8 \times 6 = 4.8$ *mi. Difference:* $7.2 - 4.8 = 2.4$ *miles.*

3. $\frac{1}{4.5} = \frac{x}{27}$. *Cross-multiply:* $4.5x = 27$, *so* $x = 6$ *cm.*

4. $\frac{27}{180} = \frac{p}{100}$. *Cross-multiply:* $180p = 2700$. *Divide:* $p = 15\%$.

5. *Round to \$93. Find 10\%: \$9.30. Double for 20\%:* $9.30 \times 2 = \$18.60$.

6. $\frac{|35-40|}{40} \times 100 = \frac{5}{40} \times 100 = 12.5\%$.

7. *Up 3 floors, then down 7:* $3 + (-7) = -4$. *The elevator is 4 floors below the lobby.*

8. $(-50) + 75 = 25$. *The deposit is larger than the overdraft, so the balance becomes positive.*

9. *The coefficients are* $\frac{1}{3}$ *and* $\frac{2}{3}$. *Add:* $\frac{1}{3} + \frac{2}{3} = \frac{3}{3} = 1$. *Result:* $1x = x$.

10. *Expand:* $-4a + 8 + 3a + 15 - 7$. *Combine:* $(-4a + 3a) + (8 + 15 - 7) = -a + 16$.

11. *If* $\frac{2}{3}x = 8$, *then each* $\frac{1}{3}$ *of* x *is 4. So* $x = 3 \times 4 = 12$.

12. *Subtract 7:* $3x \leq 18$. *Divide by 3:* $x \leq 6$.

13. *Closed circle means* -6 *is included, shading right means values greater than or equal to* -6: $x \geq -6$.

Find more at
ViewMath.com/IN-Grade7

14 Each square is 10 m. Length: $9 \times 10 = 90$ m. Width: $5 \times 10 = 50$ m. Perimeter: $2(90 + 50) = 2(140) = 280$ m.

15 Two equal sides with a 60° included angle gives a triangle where all angles are 60° (isosceles with 60° forces equilateral). All three sides are 5 cm.

16 AAA alone gives many, but specifying a side length fixes the size. A triangle with all 60° angles is equilateral, so all sides are 10 cm. Exactly one triangle.

17 A vertical slice through the apex of a cone produces an isosceles triangle. The base of the triangle is a diameter of the cone's base.

18 Vertical angles are always equal. They are formed by two intersecting lines, and the opposite angles are congruent.

19 Over the y-axis: negate x, keep y. $(2, 1) \to (-2, 1)$, $(6, 1) \to (-6, 1)$, $(6, 4) \to (-6, 4)$.

20 The circle's diameter equals the side of the square: $d = 10$, so $r = 5$ cm. $A = \pi r^2 = 3.14 \times 25 = 78.5$ cm^2.

21 Patio: $12 \times 8 = 96$ ft^2. Hot tub: $\pi r^2 = 3.14 \times 9 = 28.26$ ft^2. Remaining: $96 - 28.26 = 67.74$ ft^2.

22 $SA = 2(5 \times 3) + 2(5 \times 2) + 2(3 \times 2) = 30 + 20 + 12 = 62$ cm^2.

23 $V = s^3 = 5^3 = 125$ m^3.

24 $\frac{6}{50} = 12\%$. Then 12% of $3{,}000 = 360$.

25 Difference $= 10$. Using MAD of 4: $\frac{10}{4} = 2.5$. Using MAD of 3: $\frac{10}{3} \approx 3.3$. Both are greater than 2, so the difference is meaningful.

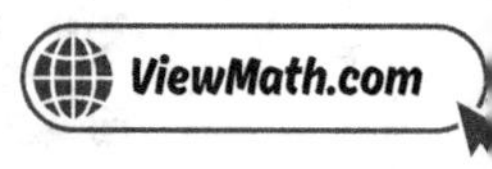

26 $Q_1 = 22$, $Q_3 = 32$. $IQR = 32 - 22 = 10$.

27 Range $= 75 - 68 = 7°$.

28 Predicted: $\frac{5}{10} = 0.5$. Experimental: $\frac{24}{40} = 0.6$. The small difference is expected from natural variation.

29 $P(tails) = \frac{1}{2}$, $P(1) = \frac{1}{6}$. $P = \frac{1}{2} \times \frac{1}{6} = \frac{1}{12}$.

30 More trials produce more reliable estimates. The experimental probability approaches the theoretical value as trials increase.

☑ Practice Test 9 — Answer Key

1 A	**2** B	**3** C	**4** D	**5** C	**6** C	**7** B	**8** $-2°C$	**9** A
10 $15x + 10$	**11** B	**12** A	**13** B	**14** B	**15** D	**16** B	**17** B	**18** B
19 A	**20** C	**21** B	**22** C	**23** B	**24** C	**25** D	**26** B	**27** C	**28** B

29 C   **30** Roll the die; let 1 and 2 represent the event (or any 2 numbers out of 6)

💡 Time to Learn! 💡

Review the explanations below, **especially for the questions you missed**.

Understanding why each answer is correct builds stronger problem-solving skills.

Tip: Circle any questions you got wrong, then read their explanation carefully.

📖 Practice Test 9 — Detailed Explanations

1. By convention, $k = \frac{y}{x} = \frac{20}{8} = 2.5$. Student B computed $\frac{x}{y}$, which is the reciprocal — not k.

2. The point $(1, 5)$ means $x = 1$ ticket and $y = \$5$, so one ticket costs $\$5$. This is the unit rate.

3. Rate: $120 \div 8 = 15$ pages/min. Time for 450: $450 \div 15 = 30$ min.

4. $\frac{60}{150} = \frac{p}{100}$. Cross-multiply: $150p = 6000$. Divide: $p = 40\%$.

5. $55 \times 1.20 = \$66.00$.

6. $\frac{|8{,}000 - 7{,}500|}{7{,}500} \times 100 = \frac{500}{7{,}500} \times 100 \approx 6.67\%$.

7. $7 - 7 = 0$. Note that $|-7| - |7| = 7 - 7 = 0$ also works, but since only one answer is expected, B is the most direct additive-inverse example. Actually D also equals 0. But B is the clearest demonstration of opposites.

8. Start at $-3°C$. Rise $4°$: $-3 + 4 = 1°C$. Drop $3°$: $1 + (-3) = -2°C$.

9. Combine $6m - 2m = 4m$. Combine $4 + 3 = 7$. Result: $4m + 7$.

10. The left section: $5 \times 3x = 15x$. The right section: $5 \times 2 = 10$. Expanded: $5(3x + 2) = 15x + 10$.

11. Multiply every term by 6: $x - 4 = 3$. Add 4: $x = 7$. Check: $\frac{7}{6} - \frac{2}{3} = \frac{7}{6} - \frac{4}{6} = \frac{3}{6} = \frac{1}{2}$ ✓.

12. Subtract 3: $2x > 8$. Divide by 2: $x > 4$.

13 $x < 5$ uses an open circle (not including 5) while $x \leq 5$ uses a closed circle (including 5). Both shade left.

14 $3.5 \times 6 = 21$ ft.

15 Two side lengths of a parallelogram do not fix the angles. The shape can be a rectangle, a rhomboid, or anything in between. Many different parallelograms are possible.

16 $7 + 10 = 17 > 15$ ✓, $7 + 15 = 22 > 10$ ✓, $10 + 15 = 25 > 7$ ✓. SSS with valid lengths gives one unique triangle.

17 A cut parallel to the base of a prism always produces a shape congruent (same size and shape) to the base. For a triangular prism, this is a triangle identical to the base.

18 $2x + 10 + 3x = 180$. Combine: $5x + 10 = 180$. Subtract 10: $5x = 170$. Divide: $x = 34$.

19 From $P(1, 2)$ to $P'(3, 0)$: x changed by $+2$ (right 2) and y changed by -2 (down 2).

20 Full circle area $= \pi r^2 = 3.14 \times 64 = 200.96$ cm^2. Quarter circle $= 200.96 \div 4 = 50.24$ cm^2.

21 Rectangle: $6 \times 4 = 24$ cm^2. Triangle: $\frac{1}{2} \times 6 \times 2 = 6$ cm^2. Total: $24 + 6 = 30$ cm^2.

22 A cube has 6 faces. $SA = 6s^2 = 6 \times 16 = 96$ cm^2.

23 $V = Bh = 24 \times 7 = 168$ cm^3.

24 The sample is the smaller group chosen from the population — the 100 students measured.

25 Class A median $= 78$, Class B median $= 65$. Class A's range (60–95) overlaps with Class B's range (50–80), but Class A is higher overall.

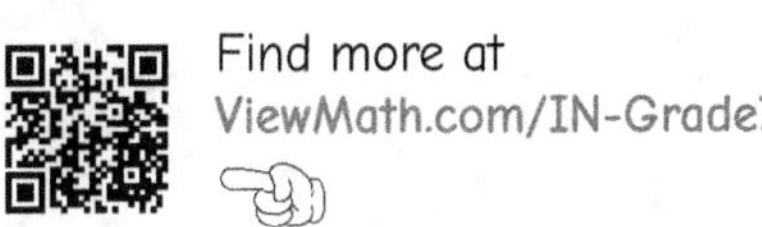

Find more at
ViewMath.com/IN-Grade7

26 Group A range $= 20$, Group B range $= 60$. Group B's data is far more spread out.

27 Team A: $3 + 4 + 2 + 5 = 14$. Team B: $2 + 5 + 3 + 4 = 14$. They tied.

28 Predicted: $\frac{1}{4} = 0.25$. Observed: $\frac{25}{80} = 0.3125$. They are close, which supports the model.

29 Only one outcome out of 8 is all tails (TTT). $P = \frac{1}{8}$.

30 Choose 2 of the 6 faces to represent the event. $P = \frac{2}{6} = \frac{1}{3}$.

✅ Practice Test 10 — Answer Key

1 B	2 B	3 C	4 C	5 C	6 Day 1 with 25% error	7 23	8 0
9 B	10 C	11 $6	12 B	13 $x < 3$; open circle at 3, shade left		14 C	15 C
16 55°	17 C	18 $a = 128°, b = 52°, c = 128°$	19 B	20 B	21 B	22 C
23 6 cm	24 C	25 B	26 Median $= 300$; IQR $= 200$	27 B	28 B	29 B	30 A

💡 Time to Learn! 💡

Review the explanations below, **especially for the questions you missed**.

Understanding why each answer is correct builds stronger problem-solving skills.

Tip: Circle any questions you got wrong, then read their explanation carefully.

📖 Practice Test 10 — Detailed Explanations

1 Use any labeled point: $k = \frac{2}{3}$ from $(3, 2)$. Check: $\frac{4}{6} = \frac{2}{3}$ ✓.

2 The point $(1, 9)$ directly gives the unit rate: 9 miles per hour. Check: $\frac{27}{3} = 9$ and $\frac{45}{5} = 9$ ✓.

3 Set up a proportion: $\frac{2}{15} = \frac{7}{d}$. Cross-multiply: $2d = 105$, so $d = 52.5$ miles.

4 Cross-multiply: $100 \times n = 250 \times 16$, which gives $100n = 4000$.

5 Each share: $70 \div 2 = \$35$. Tip: $35 \times 0.15 = \$5.25$. Total each: $35 + 5.25 = \$40.25$.

6 Day 1: $|10 - 8|/8 = 25\%$. Day 2: $|6 - 5|/5 = 20\%$. Day 3: $|8 - 10|/10 = 20\%$. Day 1 has the largest percent error at 25%.

7 The absolute value of -23 is its distance from 0, which is 23.

8 -18 and 18 are additive inverses (opposites). Their sum is always 0.

9 Combine variable terms: $3y - 5y = -2y$. Combine constants: $8 - 2 = 6$. Result: $-2y + 6$.

10 $-4 \times 2y = -8y$ and $-4 \times (-5) = +20$. A negative times a negative is positive. Result: $-8y + 20$.

11 Student $C = \frac{1}{2} \times 12 = \$6$. Check: Student $B = 31.50 - 12 - 6 = \$13.50$. Total: $12 + 13.50 + 6 = 31.50$ ✓.

12 Subtract 7: $-2x < -6$. Divide by -2 and flip: $x > 3$.

Find more at
ViewMath.com/IN-Grade7

13 Subtract 9: $-3x > -9$. Divide by -3 and flip: $x < 3$. Open circle at 3, shade left.

14 Actual dimensions: $6 \times 8 = 48$ m and $2.5 \times 8 = 20$ m. Perimeter: $2(48 + 20) = 2(68) = 136$ m.

15 AAA determines the shape but not the size. Many similar triangles of different sizes can have these angles.

16 $180° - 72° - 53° = 55°$.

17 Any cut parallel to the base of a prism gives a shape congruent to the base. A hexagonal prism has a hexagonal base, so the cross-section is a hexagon.

18 Vertical to $52°$: $b = 52°$. Supplementary to $52°$: $a = 180° - 52° = 128°$. Vertical to a: $c = 128°$.

19 Over the y-axis: negate x, keep y. $(-3, 7) \to (3, 7)$.

20 $A = \pi r^2 = \frac{22}{7} \times 49 = 154$ in^2.

21 Rectangle: $10 \times 4 = 40$ cm^2. Triangle: $\frac{1}{2} \times 10 \times 3 = 15$ cm^2. Total: $40 + 15 = 55$ cm^2.

22 $SA = 2(30 \times 20) + 2(30 \times 6) + 2(20 \times 6) = 1200 + 360 + 240 = 1{,}800$ cm^2.

23 $360 = 12 \times w \times 5 = 60w$. $w = 360 \div 60 = 6$ cm.

24 Average the two samples: $\frac{18+22}{2} = 20$ out of $50 = 40\%$. Then 40% of $1{,}000 = 400$.

25 The median is less affected by skew than the mean, so it is the best measure to compare a skewed and a symmetric distribution.

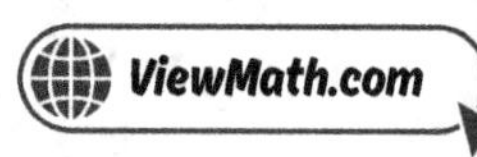

26 Median (3rd value) $= 300.$ $Q_1 = 200,$ $Q_3 = 400.$ $IQR = 400 - 200 = 200.$

27 A line graph connects data points over time, making trends easy to see.

28 Red covers half the spinner (180°), while Blue and Green each cover a quarter (90° each). This is a non-uniform model.

29 Pairs with product 12: $(2,6), (6,2), (3,4), (4,3)$ — that is 4 out of 36. $P = \frac{4}{36} = \frac{1}{9}.$

30 $P(sum = 7) = \frac{6}{36} = \frac{1}{6}.$ Expected: $36 \times \frac{1}{6} = 6.$ The simulation gave 8, slightly above expected.

Well done checking your answers!

Keep practicing to strengthen your skills.

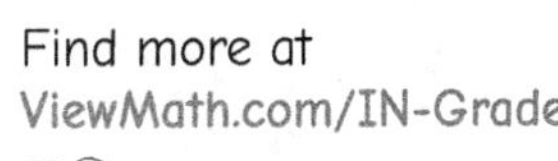

Find more at
ViewMath.com/IN-Grade7

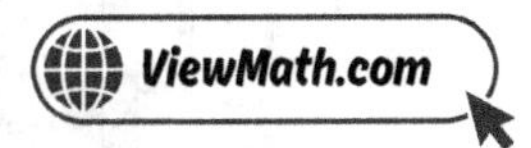

Author's Final Note

I hope you enjoyed this book as much as I enjoyed writing it. Whether you are a student working through the material, a parent supporting your child's learning, or a teacher guiding your class, I have tried to make this book as clear and engaging as possible. I hope I have succeeded. If you have any suggestions for improvement, please let me know. I would love to hear from you.

The accuracy of calculations is very important to me. We have done our best, but I also expect that I have made some minor errors. Constant improvement is the name of the game. If you find any errors, please let me know. I will fix them in the next edition.

For students: *Your learning journey does not end here. I have written a series of books to help you learn math. Make sure you browse through them. I especially recommend workbooks and practice tests to help you prepare for your exams.*

For parents: *Thank you for investing in your child's education. I encourage you to explore the companion resources available online to help support your child outside the classroom.*

For teachers: *Thank you for the invaluable work you do every day. I hope this book serves as a useful resource in your classroom. Feel free to reach out if you have suggestions or would like to discuss how best to use this book with your students.*

I also enjoy reading your reviews. If you have a moment, please leave a review on where you found this book. It will help others find this book. If you have any questions or comments, please feel free to contact me at drNazari@ViewMath.com.

And one last thing: Remember to use online resources for additional help. I recommend using the resources on https://ViewMath.com You can find video lessons, practice problems, and more. You can also use the online companion for this book to track your progress and access additional resources.

Wishing all students the best in their studies, parents every success in supporting their children, and teachers continued inspiration in their classrooms!

Dr. A. Nazari

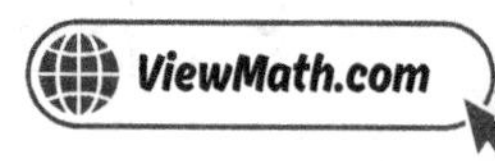

📖 *Great Job! Keep Learning with ViewMath!*

*Keep up the great work! Visit **viewmath.com/IN-Grade7** for free lessons, quizzes, and more.*

Study Guide

Workbook

Step-by-Step

3 Practice Tests

5 Practice Tests

7 Practice Tests

Find more at
ViewMath.com/IN-Grade7

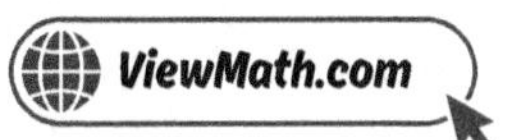